Insider Risk and Personnel Security

This textbook analyses the origins and effects of insider risk, using multiple real-life case histories to illustrate the principles, and explains how to protect organisations against the risk.

Some of the most problematic risks confronting businesses and organisations of all types stem from the actions of insiders – individuals who betray trust by behaving in potentially harmful ways. Insiders cause material damage to their employers and society, and psychological harm to the colleagues and friends they betray. Even so, many organisations do not have a systematic understanding of the nature and origins of insider risk, and relatively few have a coherent and effective system of protective security measures to defend themselves against that risk. This book describes the environmental and psychological factors that predispose some individuals to become harmful insiders, and the most common pathways by which this happens. It considers how aspects of insider risk have been altered by shifts in society, including our increasing reliance on technology and changes in working patterns. The second half of the book sets out a practical systems-based approach to personnel security – the system of defensive measures used to protect against insider risk. It draws on the best available knowledge from industry and academic research, behavioural science, and practitioner experience to explain how to make personnel security effective at managing the risk while enabling the conduct of business.

This book will be essential reading for students of risk management, security, resilience, cyber security, behavioural science, HR, leadership, and business studies, and of great interest to security practitioners.

Paul Martin, CBE, is Professor of Practice at Coventry University's London-based Protective Security Lab, a Distinguished Fellow of the Royal United Services Institute for Defence and Security Studies (RUSI), an Honorary Principal Research Fellow at Imperial College London, a member of the UK Police Science Council, and an independent adviser to various UK government entities and private sector organisations. He has a PhD from the University of Cambridge and was a Harkness Fellow at Stanford University. He is a practitioner with more than 30 years of experience in the UK national security arena.

Insider Risk and Personnel Security

An Introduction

Paul Martin

Routledge
Taylor & Francis Group

LONDON AND NEW YORK

Cover image: Getty © gremlin

First published 2024
by Routledge
4 Park Square, Milton Park, Abingdon, Oxon OX14 4RN

and by Routledge
605 Third Avenue, New York, NY 10158

Routledge is an imprint of the Taylor & Francis Group, an informa business

© 2024 Paul Martin

British Library Cataloguing-in-Publication Data
A catalogue record for this book is available from the British Library

Library of Congress Cataloging-in-Publication Data
Names: Martin, Paul, 1958 May 11– author.
Title: Insider risk and personnel security : an introduction / Paul Martin.
Description: Abingdon, Oxon ; New York, NY : Routledge, 2024 |
Includes bibliographical references and index.
Identifiers: LCCN 2023031285 (print) | LCCN 2023031286 (ebook) |
ISBN 9781032358536 (hardback) | ISBN 9781032358543 (paperback) |
ISBN 9781003329022 (ebook)
Subjects: LCSH: Risk management. | Business enterprises–Security measures.
Classification: LCC HD61 .M377 2024 (print) | LCC HD61 (ebook) |
DDC 658.15/5–dc23/eng/20230818
LC record available at https://lccn.loc.gov/2023031285
LC ebook record available at https://lccn.loc.gov/2023031286

ISBN: 978-1-032-35853-6 (hbk)
ISBN: 978-1-032-35854-3 (pbk)
ISBN: 978-1-003-32902-2 (ebk)

DOI: 10.4324/9781003329022

Typeset in Times New Roman
by Newgen Publishing UK

Disclaimer

Knowledge and practice in protective security are continually changing, as experience and research reveal more about the nature of the problems and their potential solutions. Professional practice should evolve in the light of this growing understanding and reflect changes in the threats, risks, and vulnerabilities. Leaders and practitioners should rely on their own knowledge, skills, and experience when applying any of the information, methods, or approaches described in this book. They should obtain specific professional advice that is tailored to their own particular circumstances. To the fullest extent permitted by law, the author and the publisher accept no liability for any damage or loss to persons or property, whether direct or indirect, that might arise from applying the information, methods, or approaches described herein.

Contents

Illustrations

About the author

Paul Martin, CBE, MA, PhD, FIET, is Professor of Practice at Coventry University's London-based Protective Security Lab, a Distinguished Fellow of the Royal United Services Institute for Defence and Security Studies (RUSI), an Honorary Principal Research Fellow at Imperial College London, a member of the UK Police Science Council, and an independent adviser to various UK government entities and private sector organisations. He is a practitioner with more than 30 years of experience in the UK national security arena. During a career in UK government service from 1986 to 2013 he held a variety of senior positions, including heading his organisation's personnel security function, heading the Centre for the Protection of National Infrastructure (CPNI, now NPSA), and leading national security preparations for the London 2012 Olympics. From 2013 to 2016 he was the Director of Security for the UK Parliament, with responsibility for its physical, cyber, and personnel security.

Paul was educated at the University of Cambridge, where he graduated in Natural Sciences and took a PhD in behavioural biology, and Stanford University, where he was a postdoctoral Harkness Fellow in the Department of Psychiatry and Behavioral Sciences. He subsequently lectured and researched at the University of Cambridge, and was a Fellow of Wolfson College Cambridge, before leaving academia to join government service. He is the author or co-author of several books about security and behavioural science including: *Measuring Behaviour* (Cambridge University Press, 4th edition 2021); *The Sickening Mind* (HarperCollins, 1997); *Design for a Life* (Jonathan Cape, 1999); *Counting Sheep* (HarperCollins, 2002); *Making Happy People* (Fourth Estate, 2005); *Sex, Drugs & Chocolate* (Fourth Estate, 2008); *Play, Playfulness, Creativity and Innovation* (Cambridge University Press, 2013); *Extreme* (Oxford University Press, 2014); and *The Rules of Security* (Oxford University Press, 2019).

Acknowledgements

I am very grateful to the following people for their generous help and wise advice: Andrew Glazzard, Bianca Slocombe, Charis Rice, Christina O'Kelly, David McIlhatton, Findlay Whitelaw, Jordan Giddings, Margaret Wilson, Richard Mackintosh, Sarah Austerberry, Stacy Snook, and Tara Foulsham. Two figures are reproduced by kind permission of Oxford University Press.

Introduction

We used to think of security as protecting us from bad things in the world outside. However, the worst risks can come from within. They stem from people we have trusted, and they require a different sort of security response. Human behaviour lies at the heart of these risks, making them the most interesting of all security problems, but also the most neglected.

The most corrosive security risks confronting organisations stem from the actions of insiders – individuals who abuse their trusted positions by behaving badly. Insiders inflict material damage on their employers and cause psychological injury to the colleagues and friends they betray. They are found in every type and size of organisation, from small tech start-ups to multinational corporations and government departments. Almost all the corporate risks faced by organisations have a human dimension, which means they can be affected or effected by insiders.

Many insiders know their actions are forbidden and harmful. For some, that is the whole point. Not all insiders, however, set out to cause harm. Organisations also suffer from unwitting insiders whose actions stem from recklessness, ignorance, complacency, laziness, misjudgement, or indifference. Many cyber security incidents are caused by unwitting insiders who inadvertently send data to the wrong recipients or click on attachments loaded with malware. A conventional reaction to security breaches is to blame people for being the 'weakest link'. But that misses the point. The real problem is failing to understand the risk and build the right defences against it. Unwitting insiders cause persistent low-level harm through poor performance and small acts of rebellion. However, purposeful insiders with malign intentions have the potential to cause far greater harm, and they are the primary focus of this book.

The actions of insiders are highly consequential. Insiders steal money, personal data, sensitive information, and intellectual property. They damage reputations, leak secrets, perpetrate fraud, and sabotage infrastructure. Some have killed colleagues in outbursts of violence; others have helped criminals, terrorist organisations, or hostile foreign states to do all these things. Single insider incidents have resulted in financial losses of more than a billion dollars. The commonest and costliest type of financial crime is insider fraud, with global losses running to trillions of dollars. More than half of all cyber security incidents are perpetrated by insiders.

The more capable insiders operate in the shadows and remain undetected for years. Some are never discovered. Their inside knowledge, privileged access, and authority enable them to cause more harm than the average external threat actor, while making them harder to detect. Some insiders are self-motivated, while others are manipulated or coerced by external threat actors like hostile foreign states or criminals, who can be highly proficient at exploiting human psychology.

Few organisations have a thorough understanding of insider risk or comprehensive security measures to protect them against it, creating a permissive environment in which insiders can thrive. Insider risk is less well understood and less well managed than cyber security risk. Cyber

DOI: 10.4324/9781003329022-1

security attracts much bigger budgets than personnel security, the Cinderella of protective security. Organisations spend large sums on cyber security and regularly discuss cyber risks at board level, while taking a piecemeal approach to insiders. Among those facing the starkest risks are smaller businesses that depend critically on their unique intellectual property.

People are both the sources and the instruments of the most pernicious forms of security risk. They will remain so for the foreseeable future – at least, until technology makes the next big leap into truly sentient artificial intelligence, at which point all bets are off. Insiders do more harm because they know more about their victims and have better access than the average threat actor. Fortunately, there are effective practical remedies to hand. Intelligently designed personnel security can defend organisations against insider risk.

The purpose of personnel security is to stop bad things from happening by mitigating the risk from insiders. However, good personnel security brings additional benefits by building resilience and trust. Organisations that understand and manage their insider risk are better able to avoid crises and cope with the crises they cannot avoid. The right kind of personnel security helps to build trust, which is every organisation's most valuable commodity. No organisation is safe unless it can trust its people.

The structure of this book

This book is in two parts. Part I explains the problem. Part II describes the solutions.

Chapter 1 lays the groundwork for the rest of the book by defining key terms and concepts, including insider risk and personnel security. Chapter 2 describes the many and varied types of insider behaviour and the diverse types of harm they cause, exemplified by case histories. Chapter 3 considers different types of insiders and how they can be categorised according to variable characteristics such as intentionality, autonomy, and covertness. Chapter 4 explores how insider behaviour develops through interactions between internal factors, such as personality and experience, and external factors, such as work environment and personal relationships. Finally, Chapter 5 analyses the nature of trust and its close relationship with insider risk, the characteristics that make people trustworthy, and how to judge whether a person is telling the truth.

Part II describes how a well-designed system of personnel security measures can protect organisations against insider risk. Chapter 6 sets out the basic design principles for personnel security. It explains why personnel security should take a systems approach and have a strategic purpose. Chapter 7 outlines the pre-trust personnel security measures (also known as pre-employment screening) that can be applied before a person is trusted with access to an organisation's assets. Chapter 8 describes the in-trust personnel security measures (also known as aftercare) that can be applied *after* a person has been trusted with access. Chapter 9 summarises the cross-cutting functions needed to underpin pre-trust and in-trust measures, including governance, ethics, and risk management. Chapter 10 considers ways of measuring insider risk and the effectiveness of personnel security. Finally, Chapter 11 looks at the main barriers to success in understanding and managing insider risk, including cognitive biases and a lack of systems thinking.

Language

Insider risk is complex enough without further obscuring it behind layers of opaque technical language. My aim has been to simplify wherever possible and to clarify where simplicity reaches its limits.

The problems discussed in this book are not unique to any one type of institution. The term 'organisation' is used throughout as shorthand for institutions and businesses of all types: public sector, private sector, small businesses, global corporations, universities, thinktanks, charities, government departments, intelligence agencies, police forces, and so on. The ways in which the principles and concepts are applied to protect an organisation must be tailored to its particular circumstances.

The term 'insider' refers to a person who brings heightened security risk to their organisation by exploiting, or intending to exploit, their legitimate access to the organisation's assets for unauthorised purposes. To put it another way, an insider is someone who has been trusted, but who betrays that trust by acting in potentially harmful ways. Thus, most people are not active insiders and never will be. But the few who *are* active insiders can cause immense harm.

Case histories

Real-life case histories are used throughout to illustrate the nature of insider risk and its underlying principles. The case histories are all in the public domain and the details of many have been drawn from BBC News, Sky News, *The Times*, *The Guardian*, *The Wall Street Times*, and other reputable news media. They are based on multiple sources, but single sources are cited for the sake of brevity.

Part I
Understanding insider risk

1   What is insider risk?

Reader's Guide: This chapter lays the groundwork for the rest of the book by defining key terms and concepts, including risk, insider, insider risk, and personnel security.

Before attempting to manage complex risks, it is a good idea to understand them. Security practitioners and leaders can be inclined to leap into action without pausing sufficiently to consider the nature of the risks they are trying to tackle. They are unlikely to succeed if they do not understand the risks and cannot communicate clearly about them. So, what *is* insider risk?

Insider risk is the security risk arising from the actions of insiders.

So far, so good. What, then, is meant by 'insider' and 'security risk'?

What is an insider?

There is no universally agreed definition for insider. However, according to one widely cited and entirely sensible definition:

An insider is a person who exploits, or has the intention to exploit, their legitimate access to an organisation's assets for unauthorised purposes.[1]

Thus, an insider is not just any person with access. To be an insider, they must bring heightened security risk to their organisation by exploiting, or intending to exploit, their access for purposes that might cause harm. According to this definition, most people in an organisation's workforce are *not* insiders.[2]

Viewed through a different lens, an insider is a person who has been trusted with access to an organisation's assets, and who betrays that trust by exploiting (or intending to exploit) their access for unauthorised purposes, thereby potentially causing harm. Hence, an alternative and equivalent definition is that:

An insider is a person who betrays trust by behaving in potentially harmful ways.

Other definitions are available, but they all say more or less the same thing with more words. By the same logic, and viewed through the same lens of trust:

Insider risk is the security risk arising from trusting people.

DOI: 10.4324/9781003329022-3

Trust, trustworthiness, deception, and betrayal are core concepts in personnel security. They are discussed in Chapter 5.

A person need not be a salaried employee to be an insider. Contractors, business partners, advisers, suppliers, and other third parties who are trusted with access to an organisation's assets are all potential insiders. If a high-security site is guarded by armed police officers, then they are potential insiders. Similarly, if a business outsources its IT to a supplier, then employees of that supplier are potential insiders, and so too are people in the supplier's supply chain, and so on. Having defined 'insider' we should now explain 'security risk'.

What is security risk?

Security risk is a particular type of risk.

> *Risk is the amount of harm that is likely to arise if no further mitigating action is taken.*

> <u>*Security*</u> *risks are risks that arise directly from the potentially harmful actions of threat actors such as criminals, terrorists, hostile states, and insiders.*[3]

Risk refers to possible future events that might or might not happen. Uncertainty is therefore an inherent feature of risk.

Security risk is a product of three components:

- *Threat*: the capabilities and intentions of threat actors such as criminals, terrorists, hostile states, and insiders.
- *Vulnerability*: the gaps or weaknesses in the potential victim's defences that could be exploited by a threat actor.
- *Impact*: the harm that would be caused if the risk were to materialise.

The causal chain by which threat, vulnerability, and impact combine to generate a security risk is illustrated in Figure 1.1.

Insider risk is the distinctive type of security risk arising from the actions of insiders. The concept of security risk is crucial to understanding insider risk and how to defend against it. So, let us look more closely at its three core components: threat, vulnerability, and impact.

Threat

A security threat exists when a threat actor, such as a terrorist or an insider, has both the intention and the capability to cause harm. Both ingredients are necessary. Many people harbour malign intentions but lack the capability to realise those intentions. For instance, some terrorist groups would slaughter us all if they could, but thankfully do not have the means. Conversely, some threat actors, like nation states, possess the capability to cause catastrophic harm, but usually lack the intention. Threat requires intention and capability.

The concept of threat can also be expressed as the probability (likelihood) that a threat actor will make a credible attempt to cause harm; for example, by attempting to conduct a terrorist attack or hack into a digital system. Note the word 'credible', denoting the element of capability.

National threat levels for terrorism, like those used by the UK and US governments to communicate with the public, convey threat in this same sense of likelihood of a credible *attempt*, as

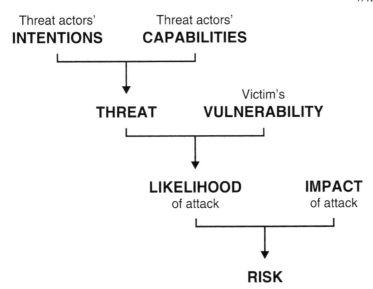

Figure 1.1 The causal chain that generates security risk[4]

Source: After Martin, 2019.

opposed to likelihood of an attack actually occurring. In practice, most terrorist plots are foiled by the pre-emptive actions of intelligence and law enforcement agencies, and consequently the number of successful attacks does not reflect the much greater extent to which credible attempts are being made. This explains why the threat level can remain high despite a prolonged absence of attacks. (The UK government definitions of threat levels are somewhat confusing because they refer to the likelihood of 'an attack' rather than a credible attempt.)

Vulnerability

Threat by itself does not make a risk. A threat actor will only succeed in causing harm if there are gaps or weaknesses in the security defences of their target, or potential victim. If the target is completely protected by impregnable defences, then the threat is likely to be deterred or thwarted. The stronger the protective security, the lower the vulnerability, and the lower the likelihood of the threat culminating in harm.

The concept of vulnerability can be expressed as the conditional probability of an attack succeeding if an attack were to be attempted. It follows that the combination of threat and vulnerability gives the likelihood of the risk materialising; for example, the likelihood of an insider defrauding their employer or stealing data.

The terminology of threat and vulnerability has its equivalents in the parallel language of crime prevention. A criminal is more likely to succeed in committing a crime if they possess a combination of motive (intention), means (capability), and opportunity. The combination of motive and means equates with threat, while opportunity equates with vulnerability. This is sometimes referred to as the COM-B model of crime, where *C*apability, *O*pportunity, and *M*otive combine to produce criminal *B*ehaviour.

Impact

We care about security risks, and spend money trying to mitigate them, because of their consequences, or impact. Thus, the third component of security risk is impact, which means the amount of harm that would occur if the risk were to materialise.

The impact of any security risk is multidimensional. An incident or attack will have many different effects, which unfold over different timescales. For example, the most immediate consequences of a major terrorist attack would be deaths and physical injuries. However, such an attack would have many other consequences as well. These are likely to include some combination of psychological injuries, damage to infrastructure, disruption to business and services, financial costs, and social and political effects. As its name suggests, the aim of terrorism is to terrorise. A major cyber attack would produce a different array of effects, such as the loss of sensitive data, disruption to business, financial costs, legal costs, loss of stakeholder confidence, and regulatory consequences. The point is that the impact of security risks, including insider risk, cannot be reduced to a simple quantitative metric, like the number of deaths or the financial cost.

Putting all three components of risk together, we see that Risk (R) is a product of Likelihood (L) and Impact (I), where Likelihood is a product of Threat (T) and Vulnerability (V):

$$R = L \times I = T \times V \times I$$

As noted before, uncertainty is an inherent feature of risk, which relates to future events that might not materialise. Measuring the *size* of a security risk is prone to further uncertainty, given the intrinsic lack of precision in any estimates of its core components of threat, vulnerability, and impact.

Comparing risks

Insider risks and other types of risks come in different shapes and sizes, so it is useful to compare them. The simplest way is by plotting the various risks on a matrix according to their likelihood and impact. An organisation might have a corporate risk matrix showing its top-level risks, such as loss of competitive advantage, reputational damage, critical staff shortages, infrastructure failure, and so on. Cyber risk usually makes the cut, but insider risk rarely appears. The organisation's internal security function (if it has one) may produce its own, more granular, matrix of security risks, such as fraud, data theft, terrorist attack, and ransomware. Insider risk should appear there, even if it is absent from the top-level corporate risk matrix.

Any one type of security risk, such as an insider incident or a terrorist attack, will encompass a range of credible scenarios that differ in their likelihood and impact. Insider incidents could vary from a harmless keyboard blunder by an unwitting employee to a catastrophic betrayal by a malign insider directed by a hostile foreign state. For a risk matrix to make sense, it must compare like with like. A general convention is to pitch each type of risk at the level of the reasonable worst-case scenario (as distinct from the most likely scenario or the absolute worst-case scenario).

The three ways to reduce security risks

Given that security risk is a product of threat, vulnerability, and impact, there are only three ways to reduce a security risk: by reducing the threat, reducing the potential victim's vulnerability, or reducing the impact (or some combination of the three).

A systematic approach to protective security involves countering each risk at every point along its causal chain (Figure 1.1) with the aim of reducing both the likelihood of the risk materialising and the impact if the risk were to materialise. This approach to risk reduction is embodied in the description of personnel security set out in Part II of this book.

The UK's national counter-terrorism strategy takes the same conceptual approach with its 'Four Ps' of Prevent, Pursue, Protect, and Prepare.[5] The Prevent agenda aims to identify and counter the underlying causes of terrorism so that individuals no longer aspire to become terrorists. This corresponds to reducing threat by reducing the intentions of potential threat actors. Pursue aims to identify and disrupt terrorists before they can conduct attacks. It tackles the symptoms by undermining the threat actors' capabilities and intentions. The Protect agenda seeks to reduce the vulnerability of potential targets. It does this by hardening the targets with protective security, making an attack plot less likely to succeed. Protect is necessary because Prevent and Pursue cannot guarantee to stop all attacks. Finally, Prepare is a backstop to deal with the consequences of attacks through measures such as incident management, business continuity planning, and disaster recovery. Prepare works mainly by reducing the impact component of the risk. It is called into play if Prevent, Pursue, and Protect fail to stop a damaging attack.

In a similar vein, a well-constructed system of personnel security measures should aim to reduce the intentions and capabilities of potential insiders, reduce the vulnerabilities of their potential targets, and reduce the impact of their hostile actions.

'Insider risk' or 'insider threat'?

As we have seen, risk and threat are different beasts. Security professionals are careful to avoid bandying the terms interchangeably, as though they mean the same. Nonetheless, people commonly refer to the security risk arising from insiders as 'insider threat'.

The terminology of 'insider threat' is prevalent in the US, reflecting the phraseology of the Presidential Executive Order on National Insider Threat Policy issued in 2012. This directed US government departments and agencies to establish 'insider threat programs' to protect classified national security information. Accordingly, the US government-funded National Insider Threat Center (NITC) defines insider threat as: 'The potential for an individual who has or had authorized access to an organization's assets to use their access, either maliciously or unintentionally, to act in a way that could negatively affect the organization'.[6] At face value, this definition appears to describe a risk, not a threat, because it refers to the amount of harm that might be caused. A more recent variant published by the US Cybersecurity and Infrastructure Security Agency (CISA) defines insider threat as 'the potential for an insider to use their authorized access or understanding of an organization to harm that organization'.[7] Again, this could be interpreted as a risk, not a threat. Others hedge their bets by referring to 'the insider threat risk', which further muddies the waters.

Some commentators are driven to use the language of 'insider threat risk' because they start with an all-encompassing definition of 'insider' as anyone with legitimate access. This makes everyone in the workforce an 'insider' and consequently requires a second label to distinguish the small minority who abuse their access to cause harm. Hence, people who could properly be referred to as insiders become 'insider threats', and the security risk arising from them becomes 'insider threat risk'. In this book, we will stick with the simpler formula of 'insider risk', and not just because it is simpler.[8]

Insider risk is a *risk*. It has three components, only one of which is threat (the others being vulnerability and impact). Apart from signalling possible confusion between threat and risk, calling it 'insider threat' detracts from the important vulnerability and impact components, both of which require their own distinct types of mitigation. Moreover, 'threat' implies a degree of intentionality that may not be present in unwitting insiders. 'Insider threat' also carries a subtle connotation of criminals, spies, or terrorists working at the behest of external threat actors, whereas many insiders are self-motivated and self-directed.

Absence of evidence of threat versus evidence of absence of risk

The conflation of risk with threat is sometimes associated with another problem, which is to confuse the absence of evidence of a threat with evidence of absence of a risk. The faulty reasoning goes like this: 'I have seen no evidence (intelligence) of a threat to this organisation and therefore it does not need more protective security'. The implicit assumption is that an apparent absence of security threat denotes an absence of security risk. The assumption is wrong at two levels.

First, an absence of evidence about security threats means just that – there is no evidence. Covert threat actors like insiders, spies, and fraudsters try to conceal their true intentions, which means that reliable information about security threats is hard to obtain. That is why the advanced capabilities of intelligence agencies are needed to discover what is really going on; and even then, the picture is always incomplete. An absence of intelligence does not signify an absence of threat, let alone an absence of risk. The only safe way to conclude that a threat actor poses no threat would be by acquiring firm evidence to that effect.

The second mistake lies in conflating threat with risk. Even if the threat component of a security risk is known to be low, the risk would still be substantial if the vulnerability and impact components are large, in which case protective security would be advisable. Suppose, for example, that reliable intelligence suggests the terrorist threat to a critical infrastructure site is low, but the site is vulnerable, and the consequences of an attack would be severe. The risk would be significant, and the site should be protected. By the same logic, an absence of evidence of a threat from insiders categorically does not prove that the insider risk is low or that there is no need to bother with personnel security. It may just show that the organisation is poor at detecting the risk.

Personnel security and vetting

If insider risk is the problem, then personnel security is the solution.

> *Personnel security is the system of protective security measures by which an organisation understands and manages insider risk.*

Part II of this book describes how to design a coherent personnel security system (with an emphasis on the word 'system').

Vague terminology can be a pitfall here as well. 'Vetting' is a widely used but ambiguous term. Some practitioners regard 'vetting' as synonymous with personnel security, in the broad sense defined above. For most people, however, 'vetting' refers only to pre-employment screening. As we shall see, there is much more to personnel security than pre-employment screening, so these two connotations of 'vetting' are very different. In the absence of evidence to the contrary, it is safer to assume that when someone talks about 'vetting', they mean only the pre-employment part of personnel security.

The 'insider risk'/'insider threat' and 'personnel security'/'vetting' confusions are not the only terminological hazards. Personnel security should not be confused with 'personal security' or 'people security'. Perso*nal* security refers to the protection of individuals against risks to their safety and security arising from other people – for instance, the risk to a media celebrity or a politician from fixated individuals. Personal security is about protecting people from other people, whereas person*nel* security is about protecting organisations from the people who work for them. The term 'people security' can mean many things, including how people contribute to security through such things as maintaining vigilance.

Security risks are dynamic and adaptive

Insider risk, in common with other types of security risk, is dynamic and adaptive. What does this mean? *Dynamic* means that the risk changes over time, sometimes rapidly. *Adaptive* means that the risk stems from intelligent threat actors who adapt their behaviour in response to the defensive actions of their potential victims. Threat actors and defenders are locked in a perpetual arms race in which both sides try to stay one step ahead of their adversary, making protective security a continuous process. As the defenders strengthen their security, so the threat actors in turn find ways of defeating or circumventing those defences, and so on.

We should never underestimate the adaptive capacity of threat actors to devise new ways of attacking us. We tend to think of creativity as an unequivocally good thing that makes the world a better place. But creativity has a dark side. Insiders, terrorists, criminals, and hostile states apply their creativity with the deliberate intent of causing harm by inventing new ways of stealing secrets, killing people, committing crimes, or undermining their adversaries. Researchers refer to this capability as malevolent creativity.[9]

Defenders can improve their chances of staying ahead in the security arms race by thinking like attackers. Security practitioners and threat actors differ enormously in their knowledge, attitudes, and experience. One of the worst mistakes a defender can make is assuming their adversaries are just like them – an error known as mirror imaging. The ability to put oneself in the mind of the adversary is a big advantage. It helps to explain why some of the best cyber security practitioners are former hackers, and why personnel security practitioners have much to learn from the specialists in police and intelligence organisations who recruit and run covert human intelligence sources.

The adaptive nature of security risks makes them different in important respects from other types of risk, such as project and programme risks, conventional fire and safety risks, many kinds of insurance and financial risks, and natural hazards like severe weather and earthquakes. These other types of risk can remain statistically relatively stable over significant periods of time. Moreover, they have materialised many times in the past, generating large volumes of actuarial data and making them amenable to quantitative analysis. For instance, there is a quantifiable risk that my house will burn down in the next year. My insurer has access to decades of actuarial data from which they can estimate this risk with considerable confidence. Moreover, the risk is unlikely to change dramatically over the next year if it is left alone. If I reduce the risk by installing smoke detectors, it will probably stay at its lower level for a while, enabling my insurer to reduce my annual premium. In contrast, the insider risk to an organisation can change overnight if one determined insider discovers a vulnerability in its security and acts on their knowledge.

The broader view of insider risk

We live in a cyber-centric world in which insider risk is often viewed through the lens of cyber security. Most organisations have cyber risk on their board-level risk register, while insider risk is relegated to the status of a facilitating factor (if, indeed, it is mentioned at all). Cyber security professionals do acknowledge that there is a human dimension to their field, and cyber security is more than just a technology problem with technology solutions. Even so, there is a widespread tendency to describe insider risk in terms of its impact on information security, as though the only thing that insiders do is to compromise digital systems and data. As we shall see in Chapter 2, the reality is very different. The manifestations of insider risk are many and varied, and its effects are certainly not restricted to the virtual world.

According to another conventional view, insiders are employees who set out to harm the organisation they work for. A broader view, which reflects the definition given earlier, is that the actions of insiders need not be aimed directly against their employer. Insiders can harm their organisation even if their actions are not deliberately directed against it, as illustrated by the two cases outlined below.

Case histories

2023: David Carrick, a former Metropolitan Police firearms officer, was jailed for dozens of rapes and sexual assaults perpetrated over decades while he was a serving officer. A senior police officer later described Carrick as 'a criminal with a warrant card'.[10]

2021: Wayne Couzens, a former Metropolitan Police firearms officer, was sentenced to life imprisonment for kidnapping, raping, and murdering a young woman, Sarah Everard.[11]

Comments

- These insider cases damaged public trust in the police, hampering their ability to perform their vital role.
- The perpetrators' criminal actions were not directed against their employer. Their victims were colleagues and members of the public. Nonetheless, they betrayed the trust of their employer and caused them significant harm, in addition to harming their victims.
- Numerous warning signs and opportunities for intervention were overlooked.
- Both men had satisfied the requirements of police personnel security. They were both firearms officers and should have been subject to even closer scrutiny. Couzens previously worked as an armed officer in two of the UK's most sensitive locations – the UK Parliament and the Sellafield nuclear storage and decommissioning site.

A striking characteristic of insiders is the diversity of their harmful actions. An insider who is determined, for whatever reason, to cause harm can do so in many different ways, such as stealing sensitive data or intellectual property, leaking damaging information, sabotaging infrastructure, or physically attacking colleagues. Case histories illustrating the diversity of insider risk are presented in Chapter 2. Nonetheless, some organisations take a narrow view of insider risk by focusing only on its most prominent symptom. So, for example, bankers chase after internal fraud, senior civil servants fret about leaking, and chief police officers search for corruption. They may not regard these problems as manifestations of the same underlying phenomenon, which is insider risk. Regulated organisations like banks have specialist functions for dealing with the symptoms of insider risk that are most relevant to their industry, like fraud and illicit trading. However, they may not fully appreciate the need for a personnel security system capable of protecting them against other insider actions.

Personnel security professionals take a broad view of insider risk and its many manifestations. They also think about the organisational and external factors that could influence the risk for better or for worse, such as internal reorganisations, bad management, remote working, pandemics, wars, ethical standards, economic crises, and declining public trust in institutions. The role of such factors in shaping insider risk is explored in Chapter 4.

Supply chain risk

Anyone with legitimate access to an organisation's physical or virtual assets is a potential insider, including full-time, part-time, and temporary employees, contractors, suppliers, business partners, and advisers. When organisations think about insider risk they tend to think mainly about their directly employed workforce. However, much of their insider risk will sit in their supply chain.

Supply chain risk is a common blind spot, possibly because of its daunting scale and complexity. For instance, consider a hypothetical business with a thousand employees. It might have a supply chain consisting of hundreds of other organisations and businesses, each of which has hundreds or thousands of employees and their own supply chains. Many of the people within that extensive supply chain will have some access to the organisation's assets – for example, because they maintain its IT systems, guard its buildings, or provide accounting services. Simple arithmetic suggests that the organisation might have more potential insiders in its supply chain than it has employees. Furthermore, it probably knows less about those people and has less control over the security regime in which they work. All of which is to say that organisations ignore supply chain risk at their peril.

Case history

2013: Edward Snowden, a former contractor for the CIA and NSA, was found to have stolen more than a million classified US government documents and leaked their military and intelligence secrets via Wikileaks and journalists, before fleeing to Russia.[12]

Comments

- A contractor can be a consequential insider.
- Insiders may exploit the trust of some colleagues to obtain highly sensitive material to which they should not have had access.
- Cyber technology can enable insiders to collect huge volumes of information – vastly more than would have been feasible in the days of paper documents.
- This insider's motives remain a matter of debate.

Discussion points

- What is the difference between insider threat and insider risk?
- Should we ditch the term 'vetting'?
- Is your organisation's security cyber-centric?
- Is your organisation's security dynamic and adaptive?
- Does your organisation understand and manage its supply chain risk?

Notes

1 This was the definition of insider used by the UK government's national technical authority (CPNI, renamed in 2023 as NPSA) for more than a decade until mid-2023.
2 NPSA changed their definition in May 2023 in line with US government terminology. Thus, everyone with current or previous authorised access became an 'insider', and the small subset that intends to or is likely to cause harm became 'insider threats'. See www.npsa.gov.uk. None of this alters the fundamental principles.
3 Martin, P. (2019). *The Rules of Security: Staying Safe in a Risky World*. Oxford: Oxford University Press.
4 Ibid.
5 HMG. (2018). *CONTEST. The United Kingdom's Strategy for Countering Terrorism*. www.gov.uk/government/publications/counter-terrorism-strategy-contest-2018
6 Gardner, C. (2018). *Five Best Practices to Combat the Insider Threat*. Carnegie Mellon University. https://apps.dtic.mil/sti/pdfs/AD1086798.pdf
7 CISA. (2023). *Defining Insider Threats*. www.cisa.gov/defining-insider-threats
8 The new NPSA definition of insider risk, with effect from May 2023, is: 'The likelihood of harm or loss to an organisation, and its subsequent impact, because of the action or inaction of an insider'. See www.npsa.gov.uk.
9 Cropley, D. H. and Cropley, A. J. (2019). Creativity and malevolence: past, present, and future. In *The Cambridge Handbook of Creativity*, 2nd edn., ed. by J. C. Kaufman and R. J. Sternberg. Cambridge: Cambridge University Press.
10 See, for example, BBC. (2023). www.bbc.co.uk/news/uk-64332586
11 See, for example, BBC. (2021). www.bbc.co.uk/news/uk-58746108
12 See, for example, BBC. (2022). www.bbc.co.uk/news/world-europe-63036991

2 Why does it matter?

Reader's Guide: This chapter describes the many and varied types of insider behaviour and the diverse sorts of harm they cause, exemplified by real-life case histories.

Types of insider actions

Insiders inflict harm in varied and imaginative ways. Their actions range from minor theft or inadvertent loss of equipment to state espionage, terrorism, and sabotage. These actions cause harm in numerous ways, such as disrupting business, damaging reputations, or killing people. When exploring the reality of insider risk, it is helpful to distinguish between insider actions and their impacts.[1] Table 2.1 lists the main types of insider actions, while Table 2.2 lists the main types of impact.

Types of impact

The impact, or harm, caused by insider actions is multidimensional, as noted in Chapter 1. Each type of insider action can have multiple consequences. In addition to the immediate effects, such as loss of money or data, the impact may include some combination of business disruption, remediation costs, erosion of stakeholder trust and confidence, legal and regulatory penalties, and reputational damage, among other things. The knock-on effects, like reputational damage or legal action, may ultimately cause more harm than the immediate impact of the insider's actions.

Other things being equal, insiders are capable of causing more harm than external threat actors because they already have access and know more about their victim. Their role may also give them authority to direct the actions of others.

The following potted case histories illustrate some of the myriad ways in which insiders have harmed those who have trusted them.

Fraud and theft

Insider fraud is the costliest and most frequent type of financial crime worldwide. The subtypes of insider fraud include stealing or skimming money, altering financial records for personal gain, bribery, abusing expenses, and benefitting from conflicts of interest. Globally, organisations are estimated to lose an average of five per cent of their revenue to insider fraud annually.[2] Of the more than 1,600 insider cases documented in the US National Insider Threat Center (NITC) database, a quarter relate to fraud in the financial services sector. The average perpetrator joined their organisation five years before starting the fraud, and the average delay between a fraud

DOI: 10.4324/9781003329022-4

Table 2.1 Types of insider actions

- Fraud (e.g. insider dealing, insider trading, abuse of payroll or expenses)
- Blackmail (e.g. obtaining money or sexual favours under threat of exposing embarrassing information)
- Theft or misuse of intellectual property (e.g. software code, designs)
- Theft or misuse of sensitive information (e.g. personal data, customer records)
- Theft of money or valuable items
- Leaking (deliberate unauthorised disclosure of sensitive or damaging information via media, social media, or other channels)
- Spilling (inadvertent disclosure of sensitive information)
- Whistleblowing (revealing actual or alleged wrongdoing)
- Facilitating unauthorised access by a third party
- Process corruption (e.g. falsifying test data or official records)
- Loss or improper disposal of assets (e.g. losing or illicitly selling IT equipment; leaving sensitive documents in a public place)
- Covert influencing (e.g. clandestinely attempting to influence government policy on behalf of a hostile foreign state)
- Exploitation of universities and other research institutions (e.g. attempting to gain access to valuable IP, people, or other organisations)
- Sabotage of physical infrastructure (e.g. damaging IT hardware)
- Sabotage of digital infrastructure (e.g. disabling IT systems, deleting data)
- Physical violence (e.g. assaulting or shooting colleagues)
- Sexual assault and harassment
- Terrorism
- Commercial espionage (covertly obtaining commercially sensitive information on behalf of a competitor)
- State espionage (covertly obtaining classified or sensitive information and passing it to a foreign state)

Table 2.2 Types of impact resulting from insider actions

- Loss or compromise of intellectual property, sensitive information, or personal data
- Loss of money or valuable items
- Loss of military, economic, or political advantage
- Erosion of democratic processes
- Loss of stakeholder trust and confidence
- Reputational damage
- Disruption of business and services
- Disruption of critical infrastructure
- Deaths and physical injuries
- Psychological injuries
- Legal consequences (e.g. compensation payments, fines, imprisonment)
- Regulatory actions (e.g. external investigations, fines, additional oversight, monitorships, tighter regulatory constraints)
- Financial costs, time costs, and opportunity costs of remediation
- Business failure

starting and being discovered was three and a half years. So-called 'low and slow' attacks, conducted covertly over a longer period, caused the most damage and took longer to discover. The commonest ways in which the insider frauds were detected were through audit checks, customer complaints, and colleagues reporting suspicions.[3]

Case histories

2022: A former employee of the Church of England was jailed for defrauding the church of £5.2M. He repeatedly diverted church funds into his own accounts over many years until he retired, spending the money on foreign travel and a lavish lifestyle. Before being employed by the church, he received a community order for theft in 1992 and was jailed in 1995 for multiple offences including 19 charges of theft.[4]

2022: The former head of Swiss bank Raiffeisen was handed a jail sentence for making millions through illicit deals while he was CEO. He had reportedly claimed £165,000 on expenses for trips to strip clubs.[5]

2021: A British paramedic was jailed for stealing thousands of pounds worth of defibrillators from his employer, the North-West Ambulance Service.[6]

2012: An employee of Lloyds Banking Group was jailed for defrauding it of £2.4M. She was head of counter-fraud and security for digital banking. She said she deserved the money for the excessively long hours she had been working.[7]

Comments

- Examples of insider fraud and insider theft.
- In some cases, an insider's previous behaviour should have rung alarm bells.

Theft of intellectual property

The most valuable asset in many organisations, apart from their people, is their intellectual property (IP): the distinctive know-how that underpins their business and gives them a competitive advantage. If your company makes perfume, for example, your IP would include the proprietary recipes for those perfumes. If your research institution makes quantum computers, your IP would be the scientific and engineering knowledge that enables you to build one that works. Organisations may seek to protect their IP with legal instruments like patents, trademarks, and copyrights. Loss of IP can seriously harm large businesses and it poses an existential threat to small companies.

An organisation's IP is rarely as clear-cut as a secret formula recorded in a document. A company's ability to sell its products often rests on the manifold skills needed to make them – for example, the precise methods by which a recipe is turned into a successful perfume, or a functioning quantum computer is built. This practical know-how resides partly in the skills and experience of key individuals. If those individuals were to leave the organisation, the know-how might disappear with them. To protect its IP, an organisation must think about its people as well as its information.

Case histories

2023: A Dutch manufacturer of semiconductor equipment reported that a former employee in China had stolen confidential company information about advanced chip-making machinery.[8]

2020: Hongjin Tan, a Chinese scientist working for a US energy company, was jailed for stealing trade secrets worth in the region of $1bn. He worked on next-generation battery technology.[9]

2020: Anthony Levandowski, a former Google engineer, was convicted of stealing Google's trade secrets about autonomous vehicle technology shortly before leaving to join its rival Uber.[10]

2007: Gary Min, a former scientist for chemical company DuPont, was jailed for stealing trade secrets valued at $400M before leaving to join a rival company.[11]

Comment

- Some insiders steal valuable intellectual property from their employer, often just before leaving the company.

Theft of nuclear material

Insiders have been involved in almost every known case in which nuclear materials such as enriched uranium or plutonium were stolen.[12]

Case history

1992: Leonid Smirnov, an engineer and trusted employee, stole 1.5 kilograms of weapons-grade highly enriched uranium from a nuclear facility in Podolsk, Russia, where he had worked for many years. He stole the uranium by repeatedly pocketing very small quantities that he knew were below the threshold for detection by the plant's auditing procedures. He did this during his colleagues' smoking breaks over a period of several months. He kept the stolen uranium in a jar on the balcony of his apartment. Smirnov intended to sell the uranium to supplement his meagre salary.[13]

Comments

- An insider may steal something valuable and dangerous that is not money or data.
- An insider may use their inside knowledge to circumvent security.

Leaking

Organisations are frequently harmed by insiders who leak sensitive or embarrassing information into the public domain. The consequences can be serious. The leakers' motives vary enormously: some believe they are acting in the public interest; others do it to settle scores or for reasons related to their feelings about themselves. Remarkably few leakers are prosecuted.

Case histories

2022: Former CIA software engineer Joshua Schulte was convicted of passing to WikiLeaks huge quantities of highly classified material detailing the CIA's hacking capabilities, in what was said to have been the largest loss of classified documents in the CIA's history. Schulte appears to have been motivated by a dispute with his employer.[14]

2021: Matt Hancock, the UK Secretary of State for Health during the Covid-19 crisis, resigned as a government minister after a newspaper published leaked pictures of him kissing his aide in his ministerial office, in breach of Covid regulations in force at the time. An official investigation found that the leaked photos were probably acquired using a mobile phone to record images from a CCTV screen.[15]

2019: Sir Kim Darroch resigned as British Ambassador to the US after a leak of classified diplomatic cables in which Darroch made disobliging comments about the then US President Donald Trump.[16]

2015: An internal auditor for UK supermarket firm Morrisons was jailed for leaking personal data on 100,000 employees. He posted the data online and sent it to newspapers. The affected employees sued the company. The leak reportedly cost Morrisons at least £2M.[17]

Comment

- Leaks can have significant consequences for international relations and national security, as well as costing a lot of money and damaging careers.

Whistleblowing

Insiders may harm their employer by revealing or alleging wrongdoing. Ethical whistleblowers do so in the public interest, often at risk to themselves. Whistleblowing about certain types of wrongdoing is protected by law.[18] Even so, people who witness wrongdoing do not always speak up because they are afraid of harming their own interests. The motivations of some self-professed whistleblowers turn out, on closer inspection, to be complicated and, in some cases, largely self-serving.

Case histories

2022: Peiter Zatko, Twitter's former head of security, testified at a US Senate hearing that Twitter had misled customers and regulators about weaknesses in its security and the number of fake accounts. The company denied the allegations.[19]

2021: Frances Haugen, a former employee of Facebook, gave evidence to US Senators and British MPs about the harm she alleged the company was causing to the mental health of teenagers. The company denied the allegations.[20]

2021: US regulator the Commodity Futures Trading Commission (CFTC) awarded nearly $200M to a whistleblower who helped them and UK regulators to investigate rigging of financial markets.[21]

Comment

- Insiders may expose or allege wrongdoing by their organisation.

Process corruption

One of many ways in which insiders cause harm is by subverting business processes – for example, by falsifying audit data required by an external regulator. This type of insider action is known as process corruption.

Case histories

2022: US metallurgist Elaine Thomas was jailed for falsifying test results for high-specification steel used in the manufacture of US Navy submarines. She falsified positive readings for tests of strength and toughness in hundreds of cases over three decades.[22]

1999: British Nuclear Fuels Ltd (BNFL) discovered that workers at their Sellafield nuclear reprocessing site had been falsifying quality assurance data since 1996. The data related to fuel for reactors. An investigation found systemic failures in management and training. BNFL had to pay compensation and take back shipments of suspect nuclear fuel from overseas. The affair cost millions in lost orders, caused severe reputational damage to BNFL, and weakened public trust and confidence in the civil nuclear industry.[23]

Comments

- Insiders can harm their employers by undermining trust in the quality of critical products, with national and international consequences.
- The reputational damage and financial impact can be severe.
- Insiders in such cases may be motivated more by complacency, laziness, or arrogance than by a conscious desire to cause harm.

Exploitation of universities and research institutions

Universities and research institutions are rich repositories of valuable knowledge and clever people who are well connected with other institutions. That makes them attractive targets for hostile foreign states, who steal their intellectual property on an industrial scale, recruit or influence their clever people, and exploit them as stepping-stones to other organisations they wish to penetrate. The easiest way to do all three is by means of insiders. Compared with penetrating high-security organisations, universities are easy meat.

China, Russia, and Iran pose the biggest security threats to western universities, although other nations are also active. Chinese intelligence operations are especially difficult to detect because they are conducted patiently over long periods and because there is no real distinction in China between private institutions and the state. Every Chinese citizen, business, and organisation is obliged under the National Intelligence Law to assist the state intelligence service if asked. Consequently, even well-intentioned students, academics, and universities could be compelled to act as instruments of the state.

Case histories

2022: A researcher at the Arctic University of Norway in Tromsø was arrested and charged with being a Russian spy. He claimed to be a Brazilian academic conducting self-funded research on security in the Arctic. Open-source investigators said they believed him to be a Russian intelligence officer.[24]

2022: A former scientist at the Norwegian University of Science and Technology was charged with illicitly exporting scientific knowledge and breaching sanctions against Iran. The scientist, who was of German and Iranian descent, had allegedly given a group of visiting Iranian scientists access to the university's research facilities and data without official consent.[25]

Comment

• These cases prompted a debate about who should be responsible for vetting foreign visitors to Norwegian universities, and indeed whether such vetting is even possible.

The insider risks to universities are not limited to classic state espionage directed at stealing their intellectual property. Universities and research institutions are also rich sources of talent and hubs for professional relationships that reach into large swathes of industry and government. Hostile foreign states exploit their easy access to universities to gain entry to harder targets like government organisations, technology companies, and multinational corporations. The two-way flow of potential insiders creates security vulnerabilities for those other organisations. For instance, a western government official or scientist might receive a flattering invitation to deliver a well-paid speech or attend an expenses-paid conference in a foreign state whose intelligence service will then start to cultivate them. While they are at it, the hosts will probably also hack the visitor's IT. Alternatively, western government officials or experts on study leave at universities in their own country may find themselves rubbing shoulders with nationals of hostile foreign states who are compelled to help their own state agencies. The eventual outcome may be the creation of an insider, who might not be consciously aware of what is happening. If the compromise is exposed, the reputational damage to the university and the people concerned can be as bad as the loss of data. Individual academics might face prosecution for breaches of export controls or United Nations sanctions.

The phenomenon is not new. Hostile foreign intelligence services have treated universities as happy hunting grounds since at least the early twentieth century. The Cambridge Spies are a well-known example. In the 1930s, Soviet intelligence recruited five Cambridge University students who went on to be prolific Soviet spies at the centre of the British establishment during World War Two and for many years thereafter: Kim Philby as a senior MI6 officer; Anthony Blunt as a wartime MI5 officer; Donald Maclean as a diplomat in the UK Foreign Office; Guy Burgess in the BBC, Foreign Office, and MI6; and John Cairncross at Bletchley Park (the wartime precursor to GCHQ) and the Foreign Office. Burgess and Maclean famously fled to Moscow in 1951, followed by Philby in 1963. The Soviet intelligence service had similar success in the US in the 1930s, recruiting the so-called Ivy League spies who later held senior positions in the US establishment.[26] These cases exemplify the Soviets' long-term approach of recruiting individuals before they had access to valuable assets.

The vulnerability of today's universities and research institutions is no mystery. They rely for their success on the free sharing of ideas and information among an eclectic mix of experts from around the world. They are by nature open and international cultures, and for good reason. Scholarship thrives on diversity and openness. Most researchers who are active in the UK are internationally mobile and more than half of the UK's research output stems from international collaborations, including with states that are strategic rivals or adversaries. Several UK universities have formal partnerships with universities in China that have links with the Chinese government and military.[27]

In response to these risks, UK government security authorities have mounted communication campaigns aimed at universities, research institutions, and industry. They outline the risks and offer guidelines for best practice in research security. The guidelines are intended to protect the integrity of international collaboration by helping universities and their industry partners to understand and manage the risks.[28]

Comparable risks arise in small-to-medium size technology companies, institutions that fund scientific research, and think-tanks. Organisations working in frontier or dual-use areas like quantum computing, quantum sensing, artificial intelligence (AI), biotechnology, robotics, and nuclear fusion possess knowledge of immense value that hostile foreign states would love to acquire – and sometimes do. The easiest way to obtain such information is usually by means of an insider. Technology companies may be more mindful of cyber security than they are of insider risk. Many start-ups initially rely on the trust between a few individuals who have known and worked with each other for years, but rapid growth and international recruitment dilute that inherent trust. Without adequate personnel security to complement their cyber and physical security, they become increasingly vulnerable to insiders.

Universities and research institutions are vital to national economic and social well-being, and their research makes huge contributions to national security and resilience. Any security response to the risks must avoid killing the golden goose with heavy-handed interventions that stifle creativity and international collaboration. Finding ways of optimising the free flow of people and ideas while managing the risk requires careful thought.

Case history

2022: Xu Yanjun, a Chinese intelligence officer, was tried and convicted in a US federal court for espionage and sentenced to 20 years in prison. He had been part of a long-running and multifaceted Chinese intelligence operation to gather sensitive information and intellectual property from western aerospace companies including GE Aviation and France's Safran Group. The operation employed a mix of cyber and human intelligence techniques, including exploiting academic institutions to gain access to western experts. Aerospace experts were invited to visit a Chinese university, with expenses and speaking fees paid by their hosts. The Chinese intelligence services used these academic visits to gain covert access to the experts' IT and to cultivate them as insiders.[29]

Comments

- The Chinese intelligence service makes use of academic exchanges to target and recruit academics and experts through whom they acquire sensitive intellectual property and access to other people of interest. The academics and experts may become conscious or unwitting insiders.

- Individuals of interest are often identified and contacted initially through the LinkedIn professional networking platform.
- The Chinese operations are long term and patient, sometimes spanning many years. They often use a mix of HUMINT (human intelligence) and cyber methods.
- Money is often utilised as a tool for enticing the putative insider and then locking them in to an increasingly clandestine relationship.

The Chinese case history summarised above illustrates the security risks associated with incautious use of social media and professional networking platforms like Facebook and LinkedIn. By advertising their personal and professional details online, people in interesting jobs make it easier for hostile foreign state agencies and other threat actors to identify them, assess their vulnerabilities, contact them, and then cultivate them. The UK National Protective Security Authority (NPSA) has run a specific communication campaign (Think Before You Link) to alert people to these risks.[30]

Sabotage

Insiders can and do break things, both in the virtual and physical worlds. They sabotage infrastructure, including cyber systems, with consequences that can be severe.

Case histories

2016: A former Citibank IT worker was jailed for sabotaging the bank's IT system after being upset by an appraisal meeting. His action cut connectivity to 90 per cent of the bank's networks in North America.[31]

2014: An insider sabotaged a nuclear power station in Belgium, causing $150M in damage. The saboteur destroyed a turbine by opening a securely locked emergency valve, thereby draining 65,000 litres of lubricating oil and causing the reactor to shut down.[32]

2000: An insider sabotaged machinery at the Sellafield UK nuclear reprocessing site by cutting cables to robotic arms used for handling radioactive waste. The facility was out of action for three days. No arrests were made.[33]

2000: A contractor repeatedly hacked into the industrial control system running a sewage system in Queensland, Australia, causing more than a million litres of raw sewage to be released into local parks, hotel grounds, and waterways. The contractor was a disgruntled former employee of the company that installed the system. He was jailed.[34]

Comments

- Insider sabotage of infrastructure can cause significant harm, including reputational damage, remediation cost, and business disruption.
- Insiders often exploit their privileged access and expert knowledge.

Physical violence

Violence is another possible manifestation of insider risk. In the US, security professionals worry about the risk of workplace shooters killing their colleagues. For centuries, trusted insiders have been assassinating political leaders, and trusted military allies have occasionally turned their weapons against their supposed friends. A more recent phenomenon is the suicidal airline pilot who kills everyone on their plane by crashing it.

It remains a notable factoid that whereas many hundreds of people have been killed over the years by trusted insiders, at the time of writing not one single person has been killed as a direct consequence of a cyber attack. None of this detracts from the importance of cyber security, where the risks are large. However, it is a salutary reminder that insider risk can have brutal consequences, and that no cyber system should be considered safe unless it is protected with physical and personnel security.

Case histories

2022: Andre Bing, a manager at a Walmart store in Virginia USA, shot dead six of his workmates and injured six others, before killing himself. Bing had worked at Walmart for more than a decade and had no previous criminal convictions. He left an incoherent note on his phone, addressed to God, declaring that his actions had not been planned and he felt he was being led by Satan.[35]

2019: French police employee Mikaël Harpon stabbed and killed four of his police colleagues and injured two others at their headquarters in central Paris, before being shot dead. Harpon was an IT specialist who had worked for the police for 16 years.[36]

2013: Aaron Alexis, a civilian contractor for the US Navy, shot dead 12 people and injured three others at the US Navy Yard in Washington DC before being killed by police. Alexis entered the facility using a valid pass. He had previously served in the US Navy and left in 2011, having been cited for misconduct on several occasions. In 2008 he was given a secret-level security clearance valid for ten years, despite being arrested for disorderly conduct the same year. He appears to have been suffering from a mental illness.[37]

2011: Royal Navy submariner Ryan Donovan was jailed for shooting two colleagues, killing one and injuring the other, on board the nuclear submarine HMS Astute while it was docked in Southampton. He had previously told a colleague he wanted to 'create a massacre'.[38]

Comments

- Examples of fatal insider violence.
- The motives are often unclear.
- Mental health problems may play a significant role in some cases.
- Workplace violence is not a uniquely US phenomenon.

Throughout history, heads of state and politicians have been assassinated at close quarters by trusted officials, bodyguards, or relatives.

Case histories

2019: Ethiopian Army General Se'are Mekonnen was shot dead by his bodyguard.

2011: Ahmed Wali Karzai, brother of the Afghan president, was assassinated by his trusted bodyguard.

1984: Indian Prime Minister Indira Gandhi was assassinated by her two trusted bodyguards.

1975: King Faisal of Saudi Arabia was shot dead at point-blank range by his nephew.

Comments

- Examples of fatal violence by trusted insiders with close access to their victims.
- The assassination of Indira Gandhi exemplified a still-common mindset that security threats come from the outside, not from within. The assassination happened after Gandhi had instigated a crackdown on Sikh separatists. The two bodyguards who killed her were Sikhs whom Gandhi had insisted on retaining despite threats to kill her.[39]

A newer form of insider violence is that of the airline pilot who commits suicide by deliberately crashing their plane. A 2022 study of crashes involving western-built civilian airliners found that murder-suicide by the pilot was the second most common cause of airline crash fatalities. Over the previous 25 years, at least six airline pilots had deliberately killed themselves, along with 543 other people on their planes.[40]

Case history

2015: Germanwings co-pilot Andreas Lubitz deliberately crashed his passenger plane into the French Alps, killing himself and 149 other people.[41]

Comments

- Pilot murder-suicide is a type of violent insider action.
- Pilots are trusted professionals. Their passengers have little choice but to trust them when they board an aeroplane.
- Lubitz had been treated for depression but concealed his mental health problems from his employer.

In military settings it is not unknown for individual members of friendly forces to suddenly turn on their supposed allies in so-called green-on-blue attacks. For example, during the long war in Afghanistan in the early twenty-first century, western soldiers from the International Security Assistance Force (ISAF) were occasionally attacked and killed by members of the allied Afghan National Security Forces who were working alongside them. These attacks became so frequent that ISAF member nations began to express reluctance about committing more troops. In early 2013, the ISAF commander described this insider problem as his top strategic threat.[42] Green-on-blue insider attacks were also a feature of the Vietnam War.

Case history

2019: Mohammed Alshamrani, a Saudi Arabian Air Force officer, shot dead three US Navy sailors and injured eight other people in an attack at the US Naval Air Station in Pensacola, Florida. Alshamrani was taking part in an aviation training programme run by the US Department of Defense for their Saudi allies. The insider attack was an act of terrorism motivated by radical Islamist ideology, for which Al Qaeda claimed responsibility.[43]

Comments

- An example of green-on-blue insider violence.
- As in many other cases, there were prior red flags about the insider's radical ideology.

Terrorism

The nature of terrorism has changed markedly over recent decades and it continues to evolve. In the UK, it is currently a complicated mix of radical Islamist, extreme right-wing, and Northern Ireland-related terrorism. The mix is different in other countries, although Islamist terrorism emanating from the likes of IS and Al Qaeda is a pervasive problem. There has also been a resurgence in state terrorism, especially by Iran, which has made many attempts in the UK to assassinate or kidnap people it regards as enemies of the regime. In 2022 the head of MI5 described Iran as the state actor that most frequently crosses into terrorism.[44] And lest we forget, Russia has conducted state terrorist attacks in the UK – notably the 2006 assassination of Alexander Litvinenko in London with a radiological weapon (Polonium-210) and the failed attempt in 2018 to kill Sergei Skripal and his daughter in Salisbury with a chemical nerve agent (Novichok).

Terrorist groups have historically tended to be cautious about using insiders to further their aims, and with good reason. Intelligence agencies have been successful over many decades in penetrating terrorist groups by recruiting covert human intelligence sources (CHIS, also known as agents) within them. Terrorists have also had good reason to be suspicious of would-be insiders who volunteer their services, as they sometimes turn out to be intelligence service agents or undercover officers. Even so, insiders have featured in several terrorist cases. The upsurge in terrorist activity by state actors like Iran makes it likely that insiders will continue to feature in future cases.

Case histories

2021: Mariam Taha Thompson, a US Defense Department contractor, was jailed for 23 years for passing classified intelligence, including the names of human sources, to a person connected with the Lebanese Hezbollah terrorist organisation.[45]

2011: Rajib Karim, a software engineer employed by British Airways, was jailed for volunteering to commit terrorist acts for Al Qaeda. Karim exploited his insider access to research ways of smuggling a bomb onto an airliner and sabotaging the airline's computer systems.[46]

2009: Humam Khalil al-Balawi, a Jordanian doctor and Al Qaeda terrorist, killed nine people including seven CIA officers and a senior Jordanian intelligence officer. Al-Balawi had previously convinced the US and Jordanian intelligence services that he was a trusted agent working for them against Al Qaeda and the Taliban. However, he secretly remained loyal to the radical Islamist cause. He lured US and Jordanian officers to a meeting at a secure CIA base in Afghanistan, where he detonated a suicide bomb.[47]

2004: A clerical worker at a Belfast hospital was jailed for collecting personal information from patient records to help Northern Irish dissident Republican terrorists to identify people to attack, including politicians, police, prison officers, and soldiers.[48]

Comments

- Examples of insider terrorism.
- Insiders can bypass physical security and information security controls.
- Insiders can dupe even cautious professional intelligence officers.

State espionage

The term espionage traditionally referred to insiders (spies) stealing secrets by covert means on behalf of an adversary. It has been described as the second oldest profession. The term now has a broader meaning and encompasses the blended use of human and technical sources. Spies do more than steal information; they may also act on behalf of an external threat actor to exert influence, sabotage infrastructure, support disinformation or active measures operations, or facilitate access to other human or technical sources of intelligence. The most active state protagonists are China, Russia, and Iran.

The espionage risk to nation states, institutions, and businesses is big and getting bigger. It affects all western nations. In 2023, for example, the head of the Australian Security Intelligence Organisation (ASIO) likened the upsurge in hostile state activity in Australia to hand-to-hand combat and said he wanted to dispel any sense that espionage is some 'romantic Cold War notion'. Similar sentiments were expressed in a joint speech by the heads of the FBI and MI5 in 2022. They placed particular emphasis on the huge and growing threat from China.[49]

Many insiders have been prosecuted in Europe for spying on behalf of state actors, principally Russia and China. An open-source study covering the period 2010–2021 found that 42 insiders had been convicted of espionage in EU and NATO member states, and a further 13 were awaiting trial. Remember that not all spies are discovered, not all spies who are discovered are prosecuted, and nations differ in their willingness to prosecute spies, all of which means that these numbers are substantial underestimates. Convictions represent only the tip of an iceberg of unknown size. As in the US, most of the spies were civilians and the great majority were men.[50]

Case histories

2023: Walter Biot, an Italian Navy captain, was jailed for 30 years for passing classified information to Russian intelligence officers in return for money. Two Russian diplomats were expelled from Italy for their part in the operation.[51]

2022: US Navy engineer Jonathan Toebbe and his wife Diana were jailed for selling sensitive information about US nuclear submarines to an undercover FBI agent posing as an agent of a foreign state.[52]

2019: CIA officer Kevin Mallory was jailed for 20 years for spying for China and exposing agents for whom he had been responsible.[53]

2019: Monica Witt, a former US Air Force linguist and counterintelligence specialist, was indicted by a US grand jury for spying on behalf of Iran. Witt had defected to Iran in 2013. She was accused of betraying US intelligence operations, including the names of agents.[54]

1961: George Blake, an MI6 officer, was sentenced to 42 years in prison for spying for the Soviet Union. Blake claimed to have betrayed more than 500 western agents, of whom more than 40 were killed. Blake escaped from jail in 1966 and fled to Russia, where he died in 2020, aged 98.[55]

Comments

- The damage to national security from insiders can include loss of access to valuable intelligence and harm to human sources (imprisonment or execution).
- Not all spies are men, though most are.

Discussion points

- What is the worst thing an insider could do to your organisation or business?
- What is the worst thing an insider could do to your country?
- How can universities deal with the foreign state insider risk while preserving openness and academic freedom?
- Is there still a role for human spies in an age of digital technology?

Notes

1 Using the new terminology published by NPSA in May 2023, an insider action would be approximately equivalent to an 'insider event', which NPSA defines as: 'The activity conducted by an insider (whether intentional or unintentional) that could result in, or has resulted in, harm or loss to the organisation'. See www.npsa.gov.uk.
2 ACFE. (2022). *Occupational Fraud 2022: A Report to the Nations*. https://legacy.acfe.com/report-to-the-nations/2022/
3 Miller, S. (2021). *Spotlight on Insider Fraud in the Financial Services Industry*. Carnegie Mellon University. https://apps.dtic.mil/sti/pdfs/AD1123958.pdf
4 Burgess, K. (2022). Jail for £5 million church fraudster. *The Times*, 20 Dec 2022.
5 BBC. (2022). www.bbc.co.uk/news/world-europe-61092952
6 BBC. (2021). www.bbc.co.uk/news/uk-england-merseyside-59267405
7 BBC. (2012). www.bbc.co.uk/news/uk-england-london-19675834
8 Robertson, J. and Koc, C. (2023). ASML stolen data came from technical repository for chip machines. *Bloomberg UK*, 15 Feb 2023.
9 DoJ. (2020). www.justice.gov/opa/pr/chinese-national-sentenced-stealing-trade-secrets-worth-1-billion
10 BBC. (2020). www.bbc.co.uk/news/world-us-canada-53659805
11 Vijayan, J. (2007). Former DuPont worker gets 18-month sentence for insider data thefts. *Computerworld*, 7 Nov 2007.
12 Hegghammer, T. and Dæhli, A. H. (2016). Insiders and outsiders. A survey of terrorist threats to nuclear facilities. In *Insider Threats*, ed. by M. Bunn and S. D. Sagan. Ithaca NY: Cornell University Press.

13 WINS. (2020). *Countering Violent Extremism and Insider Threat in the Nuclear Sector.* Version 2.0. Vienna: WINS. www.wins.org

14 BBC. (2022). www.bbc.co.uk/news/world-us-canada-62158799

15 BBC. (2022). www.bbc.co.uk/news/uk-politics-61097109

16 BBC. (2019). www.bbc.co.uk/news/uk-48939821

17 BBC. (2015). www.bbc.co.uk/news/uk-england-leeds-33566633

18 Protect. (2023). Public Interest Disclosure Act 1998 (PIDA). https://protect-advice.org.uk/pida/

19 BBC. (2022). www.bbc.co.uk/news/technology-62889754

20 BBC. (2021). www.bbc.co.uk/news/technology-59038506

21 Prentice, C. and Sims, T. (2021). Former Deutsche Bank whistleblower awarded $200 mln record payout, sources say. *Reuters*, 22 Oct 2021.

22 DoJ. (2022). www.justice.gov/usao-wdwa/pr/former-lab-director-sentenced-prison-falsifying-results-steel-testing-parts-navy-subs

23 BBC. (2000). http://news.bbc.co.uk/1/hi/uk/646230.stm

24 BBC. (2022). www.bbc.co.uk/news/world-europe-63429520

25 Myklebust, J. P. (2022). Professor in lengthy trial over Iranian scientists' visits. *University World News*, 14 Sept 2022. www.universityworldnews.com/post.php?story=20220914104835727

26 Andrew, C. (2018). *The Secret World. A History of Intelligence.* London: Allen Lane.

27 Scott, G. (2023). Universities have 'risky' ties to China. *The Times*, 22 January 2023.

28 NPSA. (2023).www.npsa.gov.uk/trusted-research

29 DoJ. (2022). www.justice.gov/opa/pr/chinese-government-intelligence-officer-sentenced-20-years-prison-espionage-crimes-attempting

30 NPSA. (2023). www.npsa.gov.uk/security-campaigns/think-you-link-tbyl-0

31 DoJ. (2016). www.justice.gov/usao-ndtx/pr/former-citibank-employee-sentenced-21-months-federal-prison-causing-intentional-damage

32 WINS. (2020). *Countering Violent Extremism and Insider Threat in the Nuclear Sector.* Version 2.0. Vienna: WINS. www.wins.org

33 BBC. (2000). http://news.bbc.co.uk/1/hi/uk/690854.stm

34 Abrams, M. and Weiss, J. (2008). *Malicious Control System Cyber Security Attack Case Study — Maroochy Water Services, Australia.* MITRE Corporation USA. www.mitre.org/sites/default/files/pdf/08_1145.pdf

35 BBC. (2022). www.bbc.co.uk/news/world-us-canada-63724716

36 Bremner, C. (2019). Anti-terror squad take on case of Mikaël Harpon, Paris police murderer. *The Times*, 5 Oct 2019.

37 DoJ. (2014). www.ojp.gov/ncjrs/virtual-library/abstracts/slipping-through-cracks-how-dc-navy-yard-shooting-exposes-flaws

38 BBC. (2011). www.bbc.co.uk/news/uk-england-14971198

39 Bunn, M. and Sagan, S. D. (2016). A worst practices guide to insider threats. In *Insider Threats*, ed. by M. Bunn and S. D. Sagan. Ithaca NY: Cornell University Press.

40 Levin, A. (2022). Murder-suicides by pilots are vexing airlines as deaths mount. *Bloomberg UK*, 13 Jun 2022.

41 BBC. (2017). www.bbc.co.uk/news/world-europe-32072220

42 Long, A. (2016). Green-on-blue violence. A first look at lessons from the insider threat in Afghanistan. In *Insider Threats*, ed. by M. Bunn and S. D. Sagan. Ithaca NY: Cornell University Press.

43 DoJ. (2020). www.justice.gov/opa/speech/attorney-general-william-p-barr-announces-findings-criminal-investigation-december-2019

44 MI5. (2022). www.mi5.gov.uk/news/director-general-ken-mccallum-gives-annual-threat-update

45 DoJ. (2021). www.justice.gov/usao-dc/pr/defense-department-linguist-sentenced-23-years-prison-transmitting-highly-sensitive

46 BBC. (2011). www.bbc.co.uk/news/uk-england-tyne-12561994

47 Warrick, J. (2012). *The Triple Agent. The al-Qaeda Mole Who Infiltrated the CIA.* NY: Doubleday.

48 BBC. (2004). http://news.bbc.co.uk/1/hi/northern_ireland/3740258.stm

49 MI5. (2022). www.mi5.gov.uk/news/speech-by-mi5-and-fbi

50 Jonsson, M. and Gustafsson, J. (2022). *Espionage by Europeans 2010–2021: A Preliminary Review of Court Cases.* Swedish Defence Research Agency (FOI) Report FOI-R-5312-SE, May 2022. www.foi.se/rest-api/report/FOI-R-5312-SE

51 Reuters. (2023). www.reuters.com/world/europe/italy-seeks-life-term-captain-accused-giving-russia-secrets-2023-03-09/

52 DoJ. (2022). www.justice.gov/opa/pr/maryland-nuclear-engineer-and-wife-sentenced-espionage-related-offenses

53 DoJ. (2019). www.justice.gov/opa/pr/former-cia-officer-sentenced-prison-espionage

54 DoJ. (2019). www.justice.gov/opa/pr/former-us-counterintelligence-agent-charged-espionage-behalf-iran-four-iranians-charged-cyber

55 See, for example: BBC. (2020). www.bbc.co.uk/news/uk-55452313

3   Who are the insiders?

Reader's Guide: This chapter characterises different types of insiders and considers how they can be categorised according to key features such as intentionality, autonomy, and covertness.

Insiders come in many different shapes and sizes, and they vary in many significant ways. There is no single profile. Moreover, there is no universally accepted taxonomic scheme for categorising insiders according to their most salient characteristics. A common but crude taxonomy divides all insiders into two types: so-called malicious insiders who deliberately cause harm; and unwitting or 'accidental' insiders who do not intend to cause harm but do so anyway. This simplistic binary distinction falls apart under closer inspection, for reasons explained below.

A taxonomy of insiders

An inspection of insider case histories shows that insiders differ from one another in several significant ways:

- *Intentionality*: the extent to which the insider consciously intends to perform illicit actions that are potentially harmful.
- *Autonomy*: the extent to which the insider's illicit actions are self-directed, as opposed to cultivated, directed, or coerced by an external threat actor.
- *Covertness*: the extent to which the insider's illicit actions are deliberately concealed, as opposed to overt and therefore easier to detect.
- *Timing*: the phase in the individual's relationship with their organisation when they first develop insider intentions (i.e. before joining, after joining, or after leaving).
- *Access*: the extent to which the insider has legitimate access to the organisation's assets and therefore the amount of harm they could cause without having to engineer additional access.

These factors all vary along a spectrum (see Table 3.1). They are not binary (either/or). Each of the case histories summarised in this book could be characterised with this taxonomy.

Two other variables that might be thought relevant do not form part of this taxonomy. They are the impact of the insider's actions and the individual's motivation. Impact is not included because the impact component of insider risk is multidimensional, as noted in Chapters 1 and 2. Insider attacks typically have multiple effects such as data loss, reputational damage, remedial costs, business disruption, and so on, which unfold over different timescales. The complicated array of effects is not easily distilled into a single label. Furthermore, insider actions often have unintended consequences, with impacts that are neither foreseen nor intended. Motivation does not feature in the taxonomy because an insider's motivation is always complicated and often

DOI: 10.4324/9781003329022-5

Table 3.1 Five key variables for categorising insiders

INTENTIONALITY:	Wholly unwitting	⟷	Fully intentional
AUTONOMY:	Externally directed	⟷	Self-directed
COVERTNESS:	Overt	⟷	Highly covert
TIMING:	Before joining	⟷	After leaving
ACCESS:	Minimal	⟷	Extensive

unknown (see Chapter 4). Very few insiders do what they do for one single reason which then becomes known to researchers.

Other variables could be added to the taxonomy, including the different types of insider actions listed in Chapter 2 (Table 2.1). However, there is a trade-off between precision and simplicity. Of the five characteristics included in this scheme, the two that probably account for much of the interesting variability are intentionality and autonomy.

What is the point of a taxonomy, you might ask. Organising insiders into categories might seem like abstract scholasticism. However, it does have a purpose. By highlighting key attributes, it helps in understanding what makes insiders tick. More importantly, a structured approach can suggest practical measures for countering the risks from different types of insiders. It is better than treating all insiders as a homogeneous blob or dividing them into two blanket categories like 'malicious' and 'accidental'. The relevance of the five variables should become more apparent in the discussion of personnel security in Part II.

Intentionality

Some insiders deliberately do things that are illicit and harmful, whereas others cause harm unintentionally. As noted above, the distinction is not binary: intentionality varies along a spectrum. At one end lie the determined insiders who purposefully break the rules and betray the trust that others have placed in them by wilfully causing harm. An archetypal example is the long-term spy who consciously betrays their nation's secrets to a hostile foreign power. At the other end of the spectrum lurks a much commoner phenomenon – the unwitting insider who has no conscious intention to cause harm but does so anyway. An archetypal example is the hapless individual who inadvertently infects their company's IT system with malware by carelessly clicking on the wrong kind of email attachment. Unwitting insiders are ubiquitous and most of us have probably been one at some point.

When assessing intentionality, it is important to distinguish between an intention to break the rules, knowing or suspecting that doing so might cause harm, and being motivated by a desire to harm the organisation. They are not the same. Intentionality is used here in the former sense – that is, deliberately doing things the insider knows are forbidden and potentially harmful, regardless of whether their purpose was to damage their organisation.

The two police insider cases outlined in Chapter 1 exemplify this distinction. Their highly intentional criminal actions were not directed against their employing organisation, but they damaged it nevertheless, as well as harming their victims. To take another example, consider a hypothetical contractor who emails a client company's financial records to their personal email address so they can work on it at home. They mistype their email address and, to their horror, find they have inadvertently sent the database to a stranger. The sensitive information later pops up on the dark web and the company suffers reputational damage, disruption, and lawsuits. The contractor's actions were intentional: they knew it was forbidden to send company information to personal email accounts because they had been told in advance. They also knew, because they were an experienced contractor, that losing the database would cause harm and probably get them sacked. Yet they sent the email because it was convenient to work from home. Their actions were not motivated by any desire to harm their client. Nonetheless, they were an intentional insider.

The expanse of spectrum between the two extremes of fully intentional and entirely unwitting insiders contains a range of interesting characters. One notch up from the wholly unwitting insiders are the reckless or incompetent insiders who knowingly do things that are illicit and potentially harmful, but for banal reasons like laziness, complacency, or ignorance. Their actions are not motivated by a wilful desire to cause harm. The hypothetical contractor described above would be an example. One notch down from the fully intentional insiders are people who sub-consciously know that what they are doing is wrong, but persuade themselves that their actions are justifiable, or that the rules do not apply to them. Our capacity for self-deception is discussed in Chapter 5.

Case histories

2021: A senior civil servant working in the UK Ministry of Defence lost classified documents at a bus stop. The documents, which contained sensitive military information, were discovered by a member of the public. An investigation found no evidence of espionage and the official was not prosecuted.[1]

2021: An error by UK Ministry of Defence personnel resulted in the exposure of personal details and photos of 250 Afghan interpreters working in Afghanistan for the British forces, putting them in danger from the Taliban, who had retaken control of the country.[2]

2021: The UK Cabinet Office was fined £500,000 for a data leak that exposed the personal details of more than a thousand recipients of official honours on a government website. The breach was thought to have resulted from human error.[3]

2008: A UK Cabinet Office official lost top-secret documents on a train in London. The documents contained sensitive information about Al Qaeda and Iraq.[4]

2008: A contractor working for the UK Home Office lost a memory stick containing unencrypted data on all 84,000 prisoners in England and Wales. The Home Office consequently terminated the contract with the contractor's management consultancy firm.[5]

Comment

• Examples of apparently unwitting insider events with significant consequences.

As noted earlier, a common binary taxonomy splits the world into 'accidental' and 'malicious' insiders. Both terms are misconceived. Arguably, there is no such thing as a truly accidental insider, and malice is in the eye of the beholder.

An unwitting insider's behaviour might be ascribed to carelessness, inattention, error, laziness, negligence, ignorance, arrogance, misjudgement, or recklessness. But it should not be described as 'accidental', as though their actions were entirely beyond their control, and they had no means of foreseeing the consequences. For similar reasons, events that were once referred to as Road Traffic Accidents (RTAs) are now called Road Traffic Collisions (RTCs). The preferred label objectively describes what happened (vehicles collided); it does not convey implicit and unproven assumptions about the underlying causes (it was an accident and not anyone's fault).

The opposite category of 'malicious' insiders is equally dubious, because it conflates intentionality with consequence. The so-called malicious insider deliberately acts in ways they know to be illicit and potentially harmful. Their actions are certainly intentional. But are they necessarily *malicious*? Consider the hypothetical contractor described earlier; they intentionally broke the rules, resulting in the unintended loss of sensitive data, but they did not do this with a malevolent intention to harm the organisation. Their actions might be described as reckless, but they were not malicious.

In some cases, the actions of an insider might seem malicious to the organisation that is harmed while being regarded by others as beneficial or public-spirited. Some highly intentional insiders could be described as virtuous people who act in the public interest. Consider the ethical whistleblower who exposes genuine wrongdoing at great personal cost, or the covert human intelligence source (agent) who provides vital intelligence on terrorism or hostile foreign state activity. The organisations whose secrets they betray might well regard them as malicious traitors, whereas the beneficiaries of their actions would have a different view. A further reason for not conflating intentionality with consequence is that insider acts can have unintended consequences, resulting in more harm than the insider might have intended.

For all these reasons, it is better not to slap the value-laden label 'malicious' on every insider whose actions are merely intentional. They might have been motivated by something other than malice, and the worst consequences might have been unintended. That said, there have been cases in which an insider does seem to have been driven by a single-minded desire to inflict harm, which could reasonably be described as malicious.

Incidentally, the practitioners of personnel security have much to learn from the professionals in intelligence and law enforcement agencies who recruit and run covert human intelligence sources (CHIS). They are experts in creating insider risk within organised crime groups, terrorist networks, and hostile foreign states. If you understand how to penetrate a secure hostile organisation by creating 'good' insiders then you should be better placed to defend your own organisation against 'bad' insiders, in much the same way as expert poachers make good gamekeepers.[6]

Commentators on cyber security sometimes assert that the risk from unwitting insiders is greater than the risk from intentional insiders. However, this assertion is not always backed by credible evidence, and it seems questionable, even in the specific context of cyber security. Unwitting insider actions are more frequent than highly intentional insider actions. However, their greater likelihood is offset by their generally lower impact. (Remember, risk is a product of likelihood and impact.) Other things being equal, a determined and intentional insider can cause more harm than an incompetent or careless unwitting insider. Evidence shows that cyber security breaches caused by intentional insiders are significantly costlier on average than those caused by unwitting insiders.[7] Moreover, unwitting insider actions are easier to detect because unwitting insiders have no strong reason to conceal them. Many intentional insider actions are never discovered or recorded, so their likelihood is under-estimated. It is therefore highly debatable whether unwitting insiders do pose a bigger risk.

Finally, a word about the taxonomic status of unwitting insiders. There was a time when what we now call unwitting insiders were not categorised as insiders at all: they were just people who made mistakes or did foolish things. They were later incorporated within the insider lexicon, largely because of their prominence in cyber security. Many cyber security breaches are caused

Table 3.2 Types of external threat actors

• Hostile foreign states (e.g. Russia, China, Iran, North Korea, Belarus)
• Terrorists
• Serious and organised criminals
• Conventional criminals
• Competitor organisations
• Ideological and single-issue activists
• Political extremists (extreme left-wing, extreme right-wing, anarchist)
• Fixated individuals
• Former employees
• Other insiders

by unwitting insiders, and their actions have attracted a lot of attention. Sceptics might argue that truly unwitting insiders are so different in many respects from intentional insiders that bracketing them together muddies the waters. That debate was long since lost, however, and unwitting insiders are now firmly entrenched within the insider family.

Autonomy

Insiders differ according to whether they are self-motivated lone actors or the products of manipulation, coercion, or direction by external threat actors, or some combination of the two. UK government research reported in 2013 that three out of four insiders are so-called self-starters, who act without any detectable involvement from external threat actors.[8] These self-starters, or self-initiated insiders, have a high degree of autonomy.

At the other end of the autonomy spectrum are insiders who are fully controlled and directed by external threat actors such as hostile foreign states or organised crime groups. Insiders who are directed by external threat actors can be extremely dangerous because they combine the privileged access and knowledge of the insider with the deep expertise and resources of the external threat actor. The archetypal example is the long-term spy. When external threat actors are involved in this way, an organisation's personnel security must defend against both insiders and outsiders.

The main types of external threat actors are listed in Table 3.2. The much greater use of IT, social media, and remote working has created more opportunities for external threat actors to identify, contact, and cultivate potential insiders. The ways in which they go about recruiting and directing insiders are discussed in Chapter 4.

A hybrid type of insider is the volunteer who spontaneously offers their services to an external threat actor and subsequently acts, more or less willingly, under their direction or control. Some notable spy cases have started this way, with an intelligence officer of one state (e.g. the UK) volunteering to spy for an adversary (e.g. Russia).

> **Case history**
>
> 1983: MI5 officer Michael Bettaney covertly volunteered his services to the Soviet KGB intelligence service in London during the height of the Cold War. He was arrested and jailed. Subsequent enquiries revealed that Bettaney had displayed many warning signs before his treachery, including erratic behaviour, heavy drinking, and minor criminality.[9]

Comments

- An example of a self-initiated and highly covert insider with extensive access who freely volunteered their services to an external threat actor.
- Hindsight revealed multiple warning signs that had been overlooked or ignored.
- The motives for his betrayal remain uncertain and might not have been apparent even to Bettaney himself.

Case history

2019: Former Israeli cabinet minister Gonen Segev was convicted of spying on behalf of Iran. He made contact with the Iranians while in Nigeria in 2012.[10]

Comment

- An insider might prefer to volunteer their services to an external threat actor while in the safer operating environment of a third country.

Covertness

Insiders vary in the covertness of their illicit actions and hence the relative ease with which they could be discovered. Covertness tends to be associated with intentionality, because highly intentional insiders usually try to conceal their illicit behaviour. Those working for sophisticated external threat actors are coached in how to avoid detection. The most successful covert insiders are never discovered. That said, some highly intentional insiders have eventually given themselves away by behaving with increasing recklessness, almost as though they unconsciously wanted to be discovered after years of concealing their activities. By contrast, genuinely unwitting insiders who believe they have nothing to hide can be remarkably overt in their behaviour. They should be relatively easy to spot.

Case history

2001: FBI officer Robert Hanssen was given 15 consecutive life sentences for spying for the Soviet Union and Russia since 1979, in what the US Department of Justice described as possibly the worst intelligence disaster in US history. Hanssen worked within the FBI's counterintelligence function. He used his insider knowledge and privileged access to avoid detection, including searching FBI records to see if he was being investigated.[11]

Comments

- A highly covert insider may remain undiscovered for decades, despite working within a high-security organisation.
- Expert inside knowledge and privileged access may be exploited to avoid detection.

Between the two extremes of deep covertness and complete overtness are insiders whose motives and personalities are complicated and whose capabilities are limited. They may attempt to conceal their illicit actions but fail to cover their tracks. Or complacency may get the better of them, leading them to cut corners and give themselves away. Even highly intentional and covert insiders make mistakes.

Timing

Insiders become insiders at different points during the course of their relationship with the host organisation. The transition to would-be insider may happen before they start working for the organisation, during the course of their engagement, in the period just before departure, or even after they have left.

Most insiders develop their insider intentions *after* they have been brought on board. The evidence from known cases shows that only a small minority join an organisation with a pre-existing intention to conduct insider actions. For example, a UK government study by CPNI (now NPSA) dating from 2013 found that only six per cent of known cases involved deliberate infiltration.[12] Insider intentions and behaviour usually develop later, emerging from a complex mix of personal predispositions, experience, circumstances, and opportunity. The developmental path by which individuals become insiders is discussed in Chapter 4.

The archetypal insider is someone who has worked for their organisation for a few years. They may also be active as an insider for quite some time. The 2013 CPNI study found that the duration of insider activity ranged from less than six months to several years.[13] In the case of insider fraud, the more senior and long serving the perpetrator, the larger the losses and the longer the fraud takes to detect, on average.[14]

Clearly, insiders who become insiders after joining are less likely to be discovered by pre-employment screening. Therefore, a hefty personnel security effort should also be directed at the existing workforce, as discussed in Chapter 8.

The third main phase of the insider life cycle is the period immediately before departure. This exit phase begins when the individual or the organisation decides that it is time for them to part company. It is a period of heightened risk for the organisation. If a person is leaving on bad terms, perhaps because they have been made redundant or dismissed for disciplinary offences, they might be sufficiently disgruntled to seek revenge. They might, for example, copy sensitive data or IP and take it with them to a rival employer. The exit process must therefore be handled carefully.

The insider risk from an individual does not disappear when they leave the organisation. Insiders can still cause harm after they have left – for example, by exploiting their inside knowledge to the detriment of their former employer. The risk is obviously worse if their former employer fails to terminate their access to its IT systems or buildings. In some cases, disgruntled former employees have been able to log on to their old accounts months after leaving, making it easy for them to cause trouble.

Case history

2010: Nathaniel Nicholson, son of convicted spy Harold Nicholson, pleaded guilty to helping his father communicate covertly with Russian intelligence from prison by smuggling out messages written on scraps of paper. Harold Nicholson, a former CIA officer, had been jailed for 23 years in 1997 for spying for Russia. Nathaniel was helping his father in his attempts to collect his 'pension' from the Russians.[15]

Comment

- An insider may continue to cause harm years after being convicted and imprisoned, drawing on their prior knowledge.

Case history

1985: Former US government official Larry Wu-tai Chin was arrested in the US and charged with spying for China. He was subsequently convicted. Chin was a naturalised US citizen who had worked as a linguist for various US government organisations, including the US Army and CIA, for 37 years. Even after retiring, he continued to pass classified documents and intelligence to the Chinese. He was only discovered after a senior Chinese intelligence officer defected to the US and exposed him. Chin killed himself in 1986, on the day he was due to be sentenced to a lengthy prison term.[16]

Comments

- A highly covert insider may remain undiscovered for decades.
- An insider may continue to cause harm after leaving their job and retiring.
- Spies may be discovered only following a defection from the other side.

Access

A fifth variable in the taxonomy of insiders is access: the extent to which the person's role affords them legitimate access to the organisation's assets. Individuals with extensive access, like senior managers, IT systems administrators, or security personnel, pose a bigger risk than those with minimal access, other things being equal. And remember that those with extensive access might include contractors or suppliers.

Note that the definition here refers to *legitimate* access – the access rights that come with the job, giving an insider a free run at the associated assets. This would be just the starting point for a determined insider, who would look for ways of extending their access beyond its legitimate boundaries. Insiders break rules. Think of Edward Snowden, for example. He took the classified data to which he had legitimate access and went far beyond that (see Chapter 1). When estimating the harm an insider could do, it is prudent to assume that they have more access than their official job description would imply.

The working definition of access in this context could be expanded to include authority – that is, the extent to which the person's role gives them legitimate authority to direct the actions of others or to perform functions like altering system software or approving financial transactions. Senior managers, IT systems administrators, and security personnel would score highly on this aspect of access as well.

More sophisticated personnel security systems take account of the relationship between access and insider risk. They do this by identifying the high-risk (high-access) roles and subjecting them to more security scrutiny. This methodology, known as role-based security, is described in Chapter 8.

Case history

2022: Omri Goren Gorochovsky, who worked as housekeeper for an Israeli defence minister, was jailed for offering to spy on his boss for an Iranian-linked group.[17]

Comment

- An insider in a junior position may be able to facilitate access to someone with much more access.

Discussion points

- Can an insider ever be truly unwitting?
- What might be different about the insiders who don't get caught?
- Are all insiders bad people?
- Should people with high levels of security clearance be paid more because they are more trustworthy?
- How are extended reality and the metaverse affecting insider risk?

Notes

1 Brown, L. (2021). Civil servant who left files at bus stop was set to be ambassador. *The Times*, 4 Aug 2021.
2 BBC. (2021). www.bbc.co.uk/news/uk-58629592
3 BBC. (2019). www.bbc.co.uk/news/uk-50929543
4 BBC. (2008). http://news.bbc.co.uk/1/hi/uk/7449255.stm
5 BBC. (2008). http://news.bbc.co.uk/1/hi/uk/7575766.stm
6 Furnham, A. and Taylor, J. (2022). *The Psychology of Spies and Spying*. Market Harborough UK: Matador.
7 Ponemon. (2022). *2022 Cost of Insider Threats Global Report*. Ponemon Institute.
8 CPNI. (2013). *CPNI Insider Data Collection Study. Report of Main Findings*. April 2013. www.npsa. gov.uk/system/files/documents/63/29/insider-data-collection-study-report-of-main-findings.pdf
9 Andrew, C. (2009). *The Defence of the Realm. The Authorized History of MI5*. London: Allen Lane.
10 BBC. (2019). www.bbc.co.uk/news/world-middle-east-46808797
11 Bunn, M. and Sagan, S. D. (2016). A worst practices guide to insider threats. In *Insider Threats*, ed. by M. Bunn and S. D. Sagan. Ithaca NY: Cornell University Press.
12 CPNI. (2013). *CPNI Insider Data Collection Study. Report of Main Findings*. April 2013. www.npsa. gov.uk/system/files/documents/63/29/insider-data-collection-study-report-of-main-findings.pdf
13 Ibid.
14 ACFE. (2022). *Occupational Fraud 2022: A Report to the Nations*. https://legacy.acfe.com/report-to-the-nations/2022/
15 *The Times*. (2010). www.thetimes.co.uk/article/spy-sent-notes-on-paper-napkins-to-russian-agents-from-jail-8g87qqkn93d
16 Engelberg, S. (1986). Spy for China found suffocated in prison, apparently a suicide. *The New York Times*, 22 Feb 1986.
17 McKernan, B. (2022). Housekeeper to Israel's defence minister jailed for offering to spy on his employer. *The Guardian*, 7 Sep 2022.

4 Why do they do it?

Reader's Guide: This chapter explores how insider behaviour develops within an individual through interactions between internal factors, such as personality and experience, and external factors, such as work environment and relationships.

Motivation

Most people are not, and never will be, an active insider (as defined in Chapter 1). What differentiates them from the small minority who *do* become active insiders? Accounts of insider cases cite a dizzying variety of putative motivators, including money, ideology, resentment, revenge, self-aggrandisement, lack of purpose, sex, curiosity, and boredom. It has been suggested, for example, that a recurrent feature of spies is an intolerable sense of personal failure.[1]

So, what *does* motivate insiders? It might seem obvious that the way to find out would be to ask them. However, there are several reasons why this is unlikely to work:

- Many insiders, especially the more capable ones, remain undiscovered. These unknown insiders might be significantly different from the insiders we know about.
- Of those insider cases that are discovered, many are not reported. Organisations are often reticent about publicising insider incidents internally, let alone externally, for fear of damaging their reputation. They may not investigate the reasons behind a case or share their learning with other organisations.
- A known insider might not cooperate with the authorities or agree to be interviewed by researchers.
- Even if an insider does agree to reveal their motives, they might not tell the truth. They might claim, for example, that they were driven by ideology when it was really about self-esteem or money. Insiders are rule-breakers, by definition, so they might just lie. Kim Philby, the MI6 officer who spied for the Soviet Union for decades, later insisted that his massive betrayal had been motivated by his private ideological convictions about the rightness of communism. But those who knew him believed it was rather more complicated than that.[2]
- An insider might have no clear insight into their own motivation, leaving them genuinely unable to articulate the reasons for their behaviour. This was apparently the case for the MI5 traitor Michael Bettaney and the workplace shooter Andre Bing described in Chapter 2.
- Evidence from known cases shows that the motivations of insiders are complicated and differ widely between individuals. Very few insiders are motivated by only one thing, such as money, and each insider is different.

DOI: 10.4324/9781003329022-6

- An insider's motivations change over time. The reasons why they originally became an active insider may be different from the reasons why they persist. The passage of time can make their initial motivations hard to recall accurately, and they may come to believe a personal narrative that is not objectively true.
- Insiders who are cultivated and directed by external threat actors like hostile foreign states may be partially or wholly unaware of how they have been manipulated. Some insiders do not know that they are insiders. Others are partially aware of what is happening but prefer to delude themselves into believing otherwise.

For all these reasons, asking an insider why they are (or were) an insider may not produce credible answers. A researcher reading about a case, perhaps years later, cannot reliably infer what was going on in the insider's mind. Indeed, there might be limited value in even attempting to analyse the motivations of individual insiders. Some accounts disappear down a rabbit-hole of motivational analysis, despite a lack of solid evidence, making an already complex problem even more impenetrable. The temptation to speculate about what motivates individual insiders should be resisted. Fortunately, there is a better way.

How (insider) behaviour develops

No credible scientist believes that human behaviour can be explained by a single causal factor, such as a gene or an experience, or that behaviour can be neatly divided into two categories: either 'innate' or 'learned'. Scientific research over many decades has consistently shown that the behaviour of humans and other animals emerges during their lifetimes through a developmental process in which genetic factors, environmental factors, and experiences interact to generate the observed behaviours.[3] What is true for human behaviour in general is true for insider behaviour in particular.

To understand the origins of an insider's behaviour, we must look beyond single causes or explanations that consider only the characteristics of the individual. Any credible explanation must encompass internal factors, external factors, and their interactions. Examples of internal factors include the individual's genetic makeup, psychological predispositions, personality traits, beliefs, knowledge, skills, emotions, mental health, and experience. External factors include their home and work environments, relationships, economic circumstances and, in some cases, the influence of external threat actors. Insider actions do not spring fully formed from the insider's beliefs, personality, or workplace experience: rather, they emerge from interactions between these and other factors. To put it another way, insiders are not born: they are made.

Thinking about other people's behaviour in terms of interactions between internal and external factors does not come naturally, partly owing to a universal psychological predisposition known as fundamental attribution bias. This is our tendency to attribute observed behaviours to internal characteristics of the other person, as though their circumstances were somehow irrelevant. However, when it comes to explaining our *own* behaviour, we err in the opposite direction, blaming our flaws and foibles on external factors. So, for example, if we see a stranger spill their coffee, we tut at their clumsiness, whereas when we do the same, we blame the slippery floor or the stressful meeting. Psychological predispositions and their implications for security are discussed in Chapter 11.

Let us look now at some of the internal and external factors that are thought to be most relevant to the development of insider behaviour.

Internal factors

Personality

Personality is the characteristic style in which an individual interacts with the world around them or, to put it another way, their consistent tendency to behave in distinctive ways. (By analogy, the culture of an organisation may be thought of as its personality.) In more technical terms, personality is a set of individual characteristics that account statistically for consistent patterns of feeling, thinking, and behaving. Psychologists argue about whether personality traits *explain* our behaviour, as opposed to being descriptive labels for statistical clusters. Either way, they are helpful concepts when analysing the origins of insider behaviour.

Psychologists have identified and measured many different characteristics that combine to form personality. One of the best-known models of personality arranges these characteristics into five broad categories, known as the Big Five. They are: *Extraversion* (enthusiastic and outgoing as opposed to quiet and aloof); *Neuroticism* (prone to anxiety and worry as opposed to emotionally stable); *Conscientiousness* (disciplined, organised, and self-controlled as opposed to careless and spontaneous); *Agreeableness* (empathetic, cooperative, and trusting as opposed to hostile and uncooperative); and *Openness* (creative, imaginative, and eccentric as opposed to conventional and practical).[4] Genetic differences between individuals feed into these personality traits in complex and subtle ways, but here is not the place to analyse them.

Of the Big Five personality traits, low Agreeableness and low Extraversion are found to be related to heightened insider risk.[5] Nested in among the Big Five are more tightly defined personality traits that are known or suspected to contribute to the development of insiders. They include Sensation Seeking: the tendency to seek varied, novel, and intense sensations and experiences and the willingness to take risks for the sake of such experiences; and Impulsivity: the tendency to act without thinking or considering the consequences. Gullibility and poor critical thinking skills are likely to contribute when insiders are manipulated by others. Mental health problems also play a role in some cases.

The Dark Triad

When it comes to identifying potential insiders, the most revealing personality traits are the so-called Dark Triad:

- *Psychopathy* (low empathy, manipulative, impulsive, disdainful, capable of cruelty)
- *Narcissism* (vain, egocentric, grandiose, sense of entitlement, excessively self-loving, sensitive to criticism, needful of admiration)
- *Machiavellianism* (exploitative, cynically manipulative of others, charming, socially skilful, ambitious)

Individuals who score highly on these traits are characterised by lack of empathy, malevolence, self-promotion, emotional coldness, duplicity, and a strong tendency to manipulate others.[6] They can nonetheless be charming and persuasive. Many spies and charismatic leaders have been described as charming but manipulative sociopaths. The Dark Triad traits are also found to be associated with malevolent creativity – the application of creativity with the deliberate intent to cause harm (as mentioned in Chapter 1).[7]

Narcissism appears to be especially relevant to insider risk.[8] Narcissists typically have a stronger propensity for anger, and they are more likely to seek revenge when provoked. Revenge can be a powerful driver of insider risk. Psychologists believe the capacity for revenge evolved in humans and other species as a deterrent against threats to the individual's well-being.[9] There is reasonably good evidence linking the Dark Triad traits to heightened insider risk, although it is uncertain whether the observed associations are directly causal as opposed to correlational.[10]

A more exotic trait is the psychological disposition to hoard possessions to an extent that becomes pathological. The hoarding tendency can express itself as digital hoarding, whereby individuals feel compelled to hoard digital items and data such as emails, photos, and files.[11] Digital hoarding could contribute to insider risk if the hoarder's compulsion to accumulate data drives them to breach security policies. If their behaviour then results in the loss or compromise of the hoarded data, the impact will be greater because of the sheer amount of data held. Risk-averse organisations with audit and compliance cultures may, paradoxically, incentivise people to hoard information in case they are later required to account for their actions.

Case history

A longstanding employee of a high-security organisation in the UK used its IT system to access highly classified documents without authority or reason. The employee, who held national security vetting clearance, hoarded the documents and shared them with some colleagues. An investigation found that the insider's actions were not motivated by a malign intention to cause harm. Colleagues knew about the individual's propensity to harvest information, but they regarded the behaviour as benign or eccentric and did not report it.[12]

Comments

- An example of data hoarding by an insider.
- The insider's actions were intentional but not malicious.
- Colleagues noticed the behaviour but chose not to report it.

Another interesting ingredient in the psychological mix is the individual's sense of their own identity, or self-concept. This construct defines who and what we think we are, and how others view us. Our sense of identity reflects our own personal combination of attributes, including our beliefs, values, physical appearance, relationships, roles, reputation, and membership of groups.[13] A person whose sense of identity is strongly affected by how they believe others view them might be more susceptible to social pressure and more motivated to seek external recognition for their achievements.

External factors

Many different aspects of an individual's current and past environments may contribute to the development of insider behaviour. At their simplest, these external factors can be divided into four broad categories:

- *Workplace factors*, such as job demands and interactions with managers and colleagues
- *Domestic and social factors*, such as relationships with family and friends
- *Societal factors*, such as economic and political issues

- *External threat actors*, in those cases where an insider is manipulated or directed by an external threat actor

As we saw in Chapter 3, most insiders become insiders after joining their organisation. Their experiences at work are often highly influential in the development of insider intentions. Factors that are thought to increase the risk include unethical leadership, incompetent management, injustice, being made to feel undervalued or insufficiently rewarded, poor organisational culture, interpersonal conflict with managers or colleagues, chronic stress, job insecurity, inadequate pay, excessive work demands, impending job loss, and exposure to malign influencers. Work also provides the potential insider with access to valuable assets and opportunities to conduct insider actions.

A persistently stressful working environment can make people feel alienated, and hence primed to become intentional insiders, or prone to making mistakes, leading perhaps to unwitting insider actions. A disruptive event such as a major reorganisation may also heighten the risk. Evidence suggests that employees are more likely to seek revenge against their employer if they work in an organisation where rule-breaking is prevalent and people are treated unfairly.[14]

The individual's domestic and social life will also influence their propensity to become an insider, in either direction. An unhappy relationship with a partner or close relative might be a predisposing factor, whereas stable and supportive relationships could act as protective factors, reducing the risk. Organisations can help to nurture some of these supportive relationships through such things as social events, family days, and flexible working.

Insider risk is further influenced by external factors in the wider environment, such as economic crises, pandemics, new technologies, social media, societal attitudes towards employment and privacy, increased dependence on outsourcing, polarisation along cultural and ideological divides, and declining public trust in institutions. The Covid-19 pandemic amplified several factors that are likely to have fostered insider risk, including psychological and financial stress and remote working. Hostile foreign states and criminals were quick to seize the opportunities. A global survey of cyber incidents involving insiders recorded a 44 per cent increase between 2020 and 2022.[15] Wars and international tensions heighten the risk of hostile foreign states recruiting insiders. Chaos and conflict are bad news for most of us, but they represent opportunities for some.

External threat actors

We saw in Chapter 3 that insiders vary in their autonomy – that is, the extent to which their actions are self-initiated and self-directed. Some insiders are cultivated, socially engineered, or coerced into becoming insiders by criminals or hostile foreign states. Their actions are then directed by the external threat actor. In some cases, the insider may be largely or wholly unaware of the extent to which they have been manipulated. The most sophisticated threat actors are highly adept at exploiting psychology to achieve their goals.

External threat actors use a variety of methods for recruiting and directing insiders. The three main methods are social engineering, material reward, and coercion. Many threat actors use a blend of all three.

Social engineering refers to a range of techniques for making the victim (and future insider) want to help the threat actor without becoming aware of the threat actor's true intentions. At its most basic, social engineering might involve phoning or emailing the victim and eliciting information by pretending to be someone in authority, such as a bank employee or tax official. Most criminal frauds are conducted through social engineering. Only after the victim's bank account has been cleared do they realise that the person they were talking to was a criminal.

More sophisticated social engineering attacks are preceded by hostile reconnaissance, in which the threat actor gathers information about their potential victims. Details such as names, job titles, work roles, email addresses, and phone numbers are particularly useful. They can be easily found on many company websites and individuals' social media sites. Personal information that seems innocuous to the individual or their employer becomes a social engineering tool in the hands of a threat actor. Threat actors know that even if their target individual has been careful not to reveal too much personal information on their social media, there is a good chance that their family and friends will have been less circumspect.

The prospect of material reward in the form of money, sex, or professional advancement can motivate some potential insiders. It might be thought that many insiders do what they do for money, but it is seldom that simple. Such cases are rare, though not unknown. More commonly, money is introduced into the relationship and becomes an increasingly important motivator. Skilful threat actors aim to make their insiders dependent on the payments and frightened by the prospect of their clandestine relationship being exposed.

Case history

2023: David Smith, a security guard at the British Embassy in Berlin, was jailed for spying for Russia. He was paid for handing classified documents to his Russian intelligence service handler. Before his arrest, Smith's colleagues had expressed concerns about his mental health. He was living beyond his means and was heard making anti-UK and pro-Russian comments. Smith told the court he had been lonely and depressed after his wife went back to Ukraine. He said he had felt undervalued and badly treated by his embassy colleagues. He was finally caught in an MI5 sting operation.[16]

Comments

- Money may play a role in an insider's motivation, but it is generally not the only factor.
- This insider's role as a security guard, though relatively junior in status terms, gave him trusted access throughout the embassy.
- Evidence revealed at the trial cast doubt on the personnel security at the embassy. There had been multiple warning signs.

The most brutal and direct way in which external threat actors create insiders is by coercion. This can take various forms, including blackmail and threats of physical violence against the individual or their family. The intelligence services of some hostile foreign states, notably Russia, still use the fabled honeytrap to blackmail individuals into cooperating. Criminal threat actors sometimes resort to physical force. In a so-called tiger kidnap, a reluctant insider is forced to commit a serious crime to prevent kidnapped family members from being harmed.

Case history

2004: An armed gang stole £26M from the Northern Bank in Belfast. They did this with the reluctant help of two bank managers, whom they coerced into cooperating by holding family members hostage. The criminals had gone to the managers' homes dressed as police officers. The bank's security systems required both managers to work together to open the vault.[17]

Comments

- Reluctant insiders may be coerced into assisting external threat actors by means of a tiger kidnap.
- Dual-key security for a bank vault is an example of a structural safeguard (see Chapter 8). In this case, it proved insufficient.

An external threat actor may start with social engineering and progress to coercion. The insider is initially cultivated in an unthreatening way. Once they are hooked, the threat actor applies pressure or outright coercion by threatening to expose the secret relationship and the misdemeanours that the insider has already been persuaded to commit.

Encapsulating the development of insider behaviour

The processes by which insider behaviour emerges from interactions between internal and external factors are undoubtedly complex. Nonetheless, the underlying concepts can be distilled into something simpler.

One of the most influential descriptions of how insider behaviour develops is the Critical Path Model (CPM).[18] It was derived by analysing evidence from known insider cases to identify factors that contribute to insider risk. The CPM distinguishes four main types of contributory factor. In chronological sequence, they are: *Personal Predispositions* (e.g. mental health problems; substance abuse; personality or social issues such as manipulation of colleagues, shyness, or bullying; a history of previous rule violations; contact with criminals or other external threat actors); *Stressors* (personal, professional, or financial factors that might exacerbate or trigger predispositions); *Concerning Behaviours* (e.g. interpersonal conflict, security violations, suspicious travel); and *Problematic Organisational Responses* (e.g. lack of attention, inadequate risk assessment or investigation, counterproductive actions like summary dismissal). In a highly compressed form, the CPM could be summarised like this:

Personal Predispositions + Stressors + Concerning Behaviours + Problematic Organisational Responses → Hostile Act

The thinking behind the CPM is that a potential insider might start with personal predispositions that make them more prone to developing insider intentions. But they only become an active insider if they are exposed to stressors like work problems, financial difficulties, or illness. If this happens, they are likely to display concerning behaviours, such as breaking rules or behaving eccentrically, before going on to commit harmful insider actions. It is during this later phase that the organisation has opportunities to spot the risk and intervene before harm is done. However, the organisation may fail to do this, or even make the problem worse, by failing to respond or intervening in ways that exacerbate the risk; for instance, by subjecting the individual to aggressive interviews and not offering to help them.

Only a small minority of people are affected by all four CPM factors. The model also recognises that protective factors may offset the risk-inducing factors. For example, supportive relationships with partner, friends, or relatives could lessen the adverse effects of a stressful working environment. Other factors may amplify the risk. For example, social media use could expose a vulnerable individual to online gambling, hurtful trolls, malign influencers, conspiracy theories, fake news, illicit relationships, and illegal content. Social media and the Internet also give would-be insiders more opportunities to cause harm – for example, by leaking sensitive information or contacting external threat actors. Personnel security should include training on the risks and safe use of social media.[19]

Known insider cases fit the CPM reasonably well, which is perhaps unsurprising as it was derived from known cases.[20] The model has the attraction of being consistent with scientific understanding of human behaviour. This makes it distinctive in a field cluttered with atheoretical models concocted bottom-up from dubious data or appeals to so-called common sense. The CPM conveys the fundamental truth that insider behaviour develops over time through interactions between internal and external factors. However, it arguably gives insufficient emphasis to the roles of chance and opportunity.

Chance and opportunity

A significant but widely undervalued element in the development of insider behaviour is chance.[21] The development of insider behaviour is not deterministic: the same set of starting conditions will not always lead to the same behavioural outcomes. People act unexpectedly and inexplicably. Sometimes we do things that do not make sense to ourselves, let alone other people.

The potential insider must have access and opportunity before they can conduct harmful insider actions. In some cases, a trigger event turns a disgruntled person into an actively malign insider. An individual could be primed with a flammable mix of psychological predispositions, money problems, and stressful work, but it might take a trigger event, like a showdown with a manager, to ignite them. Furthermore, the would-be insider must plan and prepare to some extent before they can commit harmful actions, while the organisation must repeatedly fail to notice or intervene. Each of these steps along the developmental path to insider action is subject to the randomising effects of chance.

A simple model that explicitly includes trigger events and opportunity is shown in Figure 4.1.

A caveat

Authoritative models of insider risk are hard to produce because the empirical evidence is far from comprehensive and of variable quality. Much remains unknown about the origins of insider behaviour, and current theories depend to a substantial extent on extrapolation from psychology and other fields of knowledge. Where evidence *is* derived from real insider cases, it reflects the tiny minority of individuals who become active insiders and are then caught. Most people with similar backgrounds and experiences do not become harmful insiders.

The reasons why most of us do *not* progress down the insider path are even less well understood. The situation is analogous to attempting to understand human health by studying only small samples of pathological cases. Furthermore, most insider cases that have been documented come from a relatively narrow and possibly unrepresentative range of organisations – notably, defence, security, government, and national infrastructure organisations in the US and UK. Conclusions drawn from these types of organisations might not apply equally to small businesses or charities, or to organisations in countries with different cultures and legal systems.

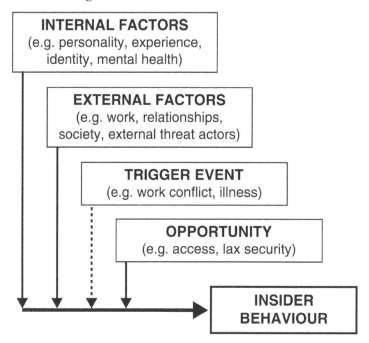

Figure 4.1 The development of insider behaviour[22]

All of which is to say that we should be careful not to push our explanations too far beyond the evidence. More research is needed.

Discussion points

- What might insider risk look like in five years' time?
- What factors might protect a person from becoming an insider?
- Is there much point in trying to work out what motivates an individual insider?
- How would you know if you were being socially engineered?
- As cyber security improves, do insider attacks become more likely?

Notes

1 Charney, D. L. (2019). *Noir White Papers on Insider Threat, Counterintelligence and Counterespionage.* www.NOIR4USA.org
2 Macintyre, B. (2014). *A Spy Among Friends: Kim Philby and the Great Betrayal.* London: Bloomsbury; Milne, T. (2014). *Kim Philby. The Unknown Story of the KGB's Master Spy.* London: Biteback.
3 Bateson, P. and Martin, P. (2000). *Design for a Life: How Behaviour Develops.* London: Vintage.
4 Nettle, D. (2007). *Personality: What Makes You the Way You Are.* Oxford: Oxford University Press.
5 Baweja, J. A., Burchett, D., and Jaros, S. L. (2019). An evaluation of the utility of expanding psychological screening to prevent insider attacks. OPA Report 2019-067. PERSEREC, US Department of Defense. https://apps.dtic.mil/sti/pdfs/AD1083812.pdf

6 Paulhus, D. L. and Williams, K. M. (2002). The Dark Triad of personality: Narcissism, Machiavellianism, and psychopathy. *Journal of Research in Personality, 36*: 556–563; Maasberg, M., Warren, J., and Beebe, N. L. (2015). The dark side of the insider: Detecting the insider threat through examination of Dark Triad personality traits. *IEEE 48th Hawaii International Conference on System Sciences*. https://doi.org/10.1109/HICSS.2015.423

7 Gao, Z. et al. (2022). Darkness within: The internal mechanism between Dark Triad and malevolent creativity. *Journal of Intelligence, 10*: 119.

8 Wilder, U. M. (2017). The psychology of espionage and leaking in the digital age. *Studies in Intelligence, 61*: 1–36.

9 Jackson, J. C., Choi, V. K., and Gelfand, M. J. (2019). Revenge: A multilevel review and synthesis. *Annual Review of Psychology, 70*: 319–345.

10 Furnham, A., Richards, S. C., and Paulhus, D. L. (2013). The Dark Triad of personality: A 10-year review. *Social and Personality Psychology Compass, 7*: 199–216.

11 Neave, N., et al. (2019). Digital hoarding behaviours: Measurement and evaluation. *Computers in Human Behaviour, 96*: 72–77.

12 Rice, C. and Searle, R. H. (2022). The enabling role of internal organizational communication in insider threat activity — evidence from a high security organization. *Management Communication Quarterly, 36*: 467–495.

13 Cheek, N. N. and Cheek, J. M. (2018). Aspects of identity: From the inner-outer metaphor to a tetrapartite model of the self. *Self and Identity, 17*: 467–482.

14 Jackson, J. C., Choi, V. K., and Gelfand, M. J. (2019). Revenge: A multilevel review and synthesis. *Annual Review of Psychology, 70*: 319–345.

15 Ponemon. (2022). *2022 Cost of Insider Threats Global Report*. Ponemon Institute.

16 BBC. (2023). www.bbc.co.uk/news/uk-64669885

17 Bunn, M. and Sagan, S. D. (2016). A worst practices guide to insider threats. In *Insider Threats*, ed. by M. Bunn and S. D. Sagan. Ithaca NY: Cornell University Press.

18 Shaw, E. and Sellers, L. (2015). Application of the critical-path method to evaluate insider risks. *Studies in Intelligence, 59*: 41–48; Costa, D. (2020). *Data-Driven Approaches to Insider Risk Mitigation*. Carnegie Mellon University. https://apps.dtic.mil/sti/pdfs/AD1110403.pdf

19 CERT. (2018). *Common Sense Guide to Mitigating Insider Threats, Sixth Edition*. Carnegie Mellon University. https://resources.sei.cmu.edu/asset_files/TechnicalReport/2019_005_001_540647.pdf

20 Shaw, E. and Sellers, L. (2015). Application of the critical-path method to evaluate insider risks. *Studies in Intelligence, 59*: 41–48.

21 In this context, 'chance' could be defined as unforeseen outcomes arising from random variations and causal factors outside the explanatory model.

22 The trigger event in Figure 4.1 has a dotted line because not every active insider requires one, whereas all insiders are influenced by internal and external factors and they all require opportunity.

5 Trust, deception, and betrayal

Reader's Guide: This chapter explores the nature of trust and its relationship with insider risk. It considers the characteristics that make people trustworthy, and ways of judging whether a person is telling the truth.

We saw in Chapter 1 that an insider can be defined as a person who betrays trust by behaving in potentially harmful ways. They have been trusted by an organisation, which gave them access to its assets, but they abuse that trust by behaving badly and potentially causing harm, whether intentionally or unwittingly.

Trust is the universal currency of insider risk and personnel security. The purpose of personnel security is to reduce insider risk by ensuring that people who have been trusted are trustworthy and remain trustworthy. As such, the concepts of trust and trustworthiness are central to personnel security.

Beyond the world of personnel security, trust is a prerequisite for a functional and happy life. Without trust, we could not maintain personal relationships, cooperate with others, engage in commerce, or sustain functioning societies. Money is a formalised abstraction of trust, embodied in cash and numbers in computers. The banking system is based on trust. When that trust breaks down, we have a credit crunch or a run on the banks. As the German sociologist Niklas Luhmann put it, a complete absence of trust would prevent one even getting up in the morning.[1]

What is trust?

The nature of trust has been debated for millennia by philosophers and writers, and investigated for decades by social scientists, psychologists, economists, lawyers, and computer scientists. Unsurprisingly, there is no universally agreed definition. That said, here is one widely quoted definition which works well for our purposes:

> *Trust is a psychological state comprising the intention to accept vulnerability based upon positive expectations of the intentions or behaviour of another.*[2]

Trusting another person or an organisation makes us vulnerable; and vulnerability is a component of risk. Thus, a central feature of trust is a willingness to put oneself or one's organisation at risk. Other definitions are available. Most are consistent with the one given above.[3]

If trust entails risk, then insider risk may be described (as it was in Chapter 1) as the security risk arising from trusting people. Trust, like risk, has an inherent element of uncertainty. If we could be absolutely certain about a person's intentions and future behaviour, we would not need to trust them. Trust is a way of managing uncertainty about the future.

DOI: 10.4324/9781003329022-7

In view of the centrality of trust, it is worth taking a closer look at the concept. A crucial distinction is the difference between trust and trustworthiness. We may trust someone or something if we believe them to be trustworthy – in other words, they possess characteristics by which we judge them to be worthy of our trust. The act of trusting is not the same as being worthy of trust, and we might decide not to trust a person even though they appear trustworthy.

Trustworthiness can be broken down into four distinct components (Table 5.1):

- *Benign intentions*: the person intends to act in your best interests (or the best interests of your organisation)
- *Integrity*: the person generally behaves towards others according to acceptable ethical standards
- *Competence*: the person has the capability to do what is expected of them
- *Consistency*: the person is reliable in consistently doing what they say they will do

The same four components would apply when judging the trustworthiness of an organisation or some other entity, such as an intelligent machine (as we shall see later).

Thus, to be confident about trusting a person, an organisation, or a robot, you must believe that they mean well by you, they behave fairly and honestly towards others, they are capable of doing what you expect of them, and they can be relied upon to do so.

Trust requires more than good intentions. To be trustworthy, a person or an organisation must also possess the competence and diligence to fulfil those intentions. Trust is therefore context dependent. It does not generalise across all situations. Trust reflects a subjective judgement about how the other party is likely to behave in a particular situation, as in: A trusts B to do X in context Y. So, for example, you might trust a colleague to perform a routine task but not trust them to give a crucial presentation to an audience of clients. Similarly, you might trust your doctor's medical advice but not place equal trust in their financial tips.

Trust is subjective and dynamic. We make subjective judgements about the trustworthiness of a person without having a full suite of objective evidence on which to base that judgement. We then update our assessment in the light of our experience of interacting with them, which may lead us to trust them more or trust them less. Trust is not simply an outcome of observing good behaviour: we trust others before having solid evidence of their good behaviour, and we withdraw our trust if they betray it by behaving badly.[4]

A broadly equivalent model of trust that is often cited in the context of insider risk is known as the ABI framework. It divides trustworthiness into three rather than four components. They are *A*bility, *B*enevolence, and *I*ntegrity.[5] The ABI components correspond to Competence, Benign Intentions, and Integrity. Other experts have found it preferable to add a fourth component which they refer to as Predictability or Consistency, and which equates with Consistency in Table 5.1.[6] This fourth component is important, as it reflects the way that trust changes over time through repeated interactions. The Consistency component captures the untrustworthy behaviour of flaky individuals who mean well but are not dependable.

We humans are predisposed to trust others, and we enjoy doing so. As the eighteenth-century philosopher David Hume wrote: 'One may create trust out of the pleasure that such sympathetic

Table 5.1 The components of trustworthiness

Benign intentions	Integrity
Competence	Consistency

feelings bring'. A biologist might argue that trusting others is pleasurable because we are social animals and our capacity for trust is an evolved characteristic. Trust facilitates cooperation, which helped our ancestors to survive and thrive in a dangerous and uncertain world. By the same token, evolution has made us highly sensitive to breaches of trust and predisposed to retaliate against those who betray us. Hence, trust is famously easier to destroy than it is to build, and people who have been betrayed may feel vengeful.

Here are some other things you need to know about trust and trustworthiness in relation to insider risk and personnel security:

- Trust is not binary. It varies along a continuum. Individuals cannot simply be categorised as either 'trustworthy' or 'not trustworthy'.
- Trust is not the same as cooperation. Two parties can cooperate without necessarily trusting one another. Legal contracts or other control mechanisms may partly substitute for trust in some relationships. However, mutual trust greatly facilitates cooperation.[7]
- Trust is not the same as loyalty, which could be described as remaining true to a person, a cause, or an organisation despite there being attractive alternatives. You can trust several organisations, but you can only be truly loyal to one.
- Trust is not the same as rapport – the mutual attraction when two people first meet and get on well. We might like someone to begin with but later discover that they cannot be trusted. Fraudsters and charismatic leaders are experts at establishing rapport. Rapport helps to establish a trusting relationship at an early stage when there is little evidence about trustworthiness.[8]
- Distrust is not simply the opposite or absence of trust. Rather, it is an active judgement not to trust. You might not trust someone about whom you know very little, but that does not necessarily mean that you actively *dis*trust them.[9]
- Trust between individuals tends to be mutually reinforcing. The more we trust someone, the more inclined they are to trust us, creating a virtuous cycle. Trust is both a cause and a consequence of good behaviour. Similarly, distrust breeds distrust, creating a vicious cycle.[10]
- Individuals differ in their general psychological disposition to trust others – a trait known as trust propensity.[11] An excessive propensity to trust can result in gullibility or blind trust, which puts the trustor at greater risk of being harmed. Conversely, a paranoid unwillingness to trust anyone makes normal life impossible.
- Other things being equal, we are inclined to judge people as more trustworthy if they resemble us in appearance or background. Experiments have shown, for example, that we are more likely to trust strangers whose faces look like ours.[12]

The value of trust

Arguably, the most valuable asset that any organisation can possess is trust – more specifically, high levels of mutual trust between individuals, between the workforce and the organisation, and between the organisation and its stakeholders. High-trust organisations trust their people, and the people trust each other, because the organisation and its people are trustworthy. As we shall see, high-trust organisations are more successful in many other ways besides having less insider risk. (The so-called zero-trust approach to cyber security is discussed in Chapter 8.)

Research has shown that interpersonal trust is one of the main contributors to optimal performance by teams. The American writer Simon Sinek has described how high-performing organisations like the US Navy SEALS have learned that trustworthiness matters even more than individual performance when it comes to building effective teams. The SEALS would rather have a highly trustworthy moderate performer than a high performer who cannot be trusted. Indeed, Sinek suggests that the quickest way of identifying an untrustworthy person (in our terms, a potential insider) is to ask team members 'Who's the asshole?'[13]

Trust has pervasive benefits that extend beyond the mitigation of insider risk. Trust reduces complexity and enables clarity of purpose. It frees people to cooperate and get on with making a difference. Trust reduces transaction costs, whereas a lack of trust generates a proliferation of lawyers, contracts, and policy documents. If you trust a person or an institution, you can reach a decision quickly and without recourse to lengthy negotiations, written agreements, and all the other tedious trappings of distrust. The social scientist Robert Putnam described reciprocal trust as a social lubricant that reduces transactional friction and makes high-trust organisations or societies more efficient, in much the same way that money is a more efficient means of trading than barter. High-trust societies are more resilient and better able to cope with economic shocks or natural disasters.[14]

Similar principles apply to organisations. Evidence shows that high-trust organisations tend to be more innovative and more resilient; they make faster decisions and adapt better to changing pressures; they waste less time and money on compliance processes; and they are more agreeable places to work, making them better able to recruit and retain talent. When there are high levels of trust, people are more willing to share information, take risks, and embrace change, making it easier to innovate and adapt. High-trust teams perform better on average, and managers can spend less time monitoring staff performance. Employees experience higher job satisfaction, reducing turnover and enhancing the organisation's reputation as a good place to work.[15] Trust is a business advantage.

Conversely, organisations or societies with low levels of trust suffer adverse consequences. The Soviet Union and East Germany during the Cold War were crippled by paranoia and suspicion. Individuals could not trust their neighbours, colleagues, or even family members, because there was a real risk that they might be informers for the secret police. East Germany harboured a huge state apparatus that compelled citizens to spy on each other, with an average of one Stasi secret police officer or informant for every 63 people in the nation.[16] Many East Germans responded by withdrawing into themselves and keeping their inner lives secret.

Low-trust organisations can expect to experience more counterproductive work behaviours, more resistance to change, less cooperation, higher turnover, more bureaucracy, and more insider actions. The shift to remote working that was accelerated by the Covid-19 pandemic put a strain on the trust between employer and workers in some organisations which insisted on monitoring their workforce to check that they were still working hard. More enlightened organisations recognise that treating everyone as though they were untrustworthy will undermine the very trust they are seeking to build. Trust has increasingly featured in the language of leadership, while the old business school penchant for admiring ruthless role models like Machiavelli has gone out of fashion.

Aspiring to create a high-trust environment does not mean blithely assuming that everyone is trustworthy unless proven otherwise. There have been countless examples of organisations ignoring warning signs about individuals who went on to cause harm, as exemplified by case histories in this book. Judgements about trustworthiness should be tested and amended if the evidence changes. As the Russian proverb says: *Doveryai, no proveryai* (Trust, but verify).

Deception

The ability to deceive others is a basic skill for any insider (with the possible exception of the truly unwitting insider who just makes an honest mistake). The most dangerous insiders are intentional and covert: their surface behaviour must conceal their real agenda. If deception is central to the insider's success, then the ability to detect deception must be central to personnel security. So, how easy is it?

Detecting deception

We are not as good at spotting lies as we like to think. Our ability to judge whether a person is telling the truth, just by talking to them, is surprisingly poor. One reason is the widespread but mistaken belief that liars give themselves away through non-verbal behaviours such as fidgeting, avoiding eye contact, swallowing, sweating, hesitation, facial expressions, shifting posture, or blinking. This is not true. It would be convenient if liars displayed the behavioural equivalent of Pinocchio's nose, but no such thing exists.

Scientists have been investigating deception for many years. The large body of evidence they have accumulated consistently shows two things: our performance at detecting lies is only slightly better than tossing a coin; and non-verbal behaviours are weak and unreliable indicators of deception.[17] Despite this, many people believe they are good at identifying lies and that looking for non-verbal cues is the way to do it. The evidence says otherwise. An authoritative meta-analysis[18] of published studies found that people's average success rate in distinguishing lies from truth was 54 per cent – in other words, barely superior to guessing. There was no difference in accuracy between experts (e.g. police officers or psychiatrists) and laypeople (e.g. students). Nonetheless, the experts were more confident in their judgements.[19] It seems we are all capable of being frequently wrong but never in doubt (as a prominent businessman was once described). The evidence also showed that people performed worse when they could see the interviewee, compared with when they could only hear them, which supports the view that non-verbal cues do not help.

To make matters worse, interviewers tend instinctively to trust people who resemble them physically, which obviously has no bearing on their actual trustworthiness. Evidence further suggests that professional investigators may be biased towards identifying deception: their experience has led them to expect people to lie, and consequently they are more inclined to judge that someone is lying. Researchers found that experienced law enforcement investigators who had received the wrong kind of training (and there has been plenty of that) were no more accurate at detecting deceit than those with little or no experience. They were, however, systematically biased towards deciding that an interviewee was lying, and more confident in their often-inaccurate judgements.[20] The training has improved in recent years, reflecting the improved understanding from research.

The evidence is clear that the best way to determine whether a person is telling the truth is to assess the *content* of what they are saying, and not be distracted by their body language. Practical techniques that can significantly improve our ability to discriminate between truth and deception when interviewing someone are described in Chapter 7.

The polygraph and other technologies

A more intrusive approach to detecting deception is the polygraph, which measures fluctuations in heart rate, breathing, blood pressure, and skin conductivity. Related technologies are capable of measuring physiological and behavioural variables at a distance, without the individual necessarily being aware. They include facial thermal imaging (to detect changes in skin temperature and blood flow), laser Doppler vibrometry (to detect changes in pulse rate and breathing), voice stress analysis, and automated analysis of non-verbal behaviour.

The polygraph and related technologies are based on the premise that the physiological and behavioural variables they record are indicators of psychological stress which, in turn, is a proxy indicator of deception. When someone lies, the theory goes, their deception is revealed by stress-related changes in their heart rate, breathing, and so on.

It would be convenient if the polygraph were a lie-detector of unerring accuracy, but it is not. Many investigations have found that, although it can bring real diagnostic value, the polygraph is not as reliable as some of its more enthusiastic proponents have claimed. When used skilfully, the polygraph can help to discriminate between truth and fiction. Part of its effectiveness lies in the theatre of its use: if the person being polygraphed is convinced of its accuracy, they may be more willing to tell the truth. A further problem is that the polygraph and comparable tools can be subverted. The physiological variables they measure can be deliberately altered by mental or physical countermeasures.[21] Some of the most damaging spies succeeded in passing polygraph tests.

Case history

1994: CIA counterintelligence officer Aldrich Ames was convicted of spying for the Soviet Union and Russia since the mid 1980s, compromising many western intelligence agents and operations. During that period Ames passed two polygraph examinations.[22]

Comments

- The polygraph failed to detect that Ames was a highly damaging spy.
- Thanks to his inside knowledge, bolstered with expert advice from his Soviet and Russian handlers, this insider managed to stay hidden within a high-security intelligence agency for many years.

Efforts have been made to detect deception or hostile intentions by monitoring brain activity with magnetic resonance imaging (MRI) or electroencephalogram (EEG) recording, though so far with only limited success. Such techniques are highly invasive, and they require the subject to be cooperative and very patient. They can detect when someone is stressed, although there are easier ways of doing that than putting someone inside a brain scanner. Moreover, there are neurobiological reasons why such methods do not readily disgorge accurate read-outs of people's covert intentions. Many different parts of the brain are active when someone is lying, and there is no tell-tale 'deception centre' to light up in an MRI scan. Talking to people is currently still the least bad way to find out what they are thinking. The picture is changing, however.

Machine learning (ML) techniques (which have been described as a posh way of doing statistics very quickly) are rapidly getting better at deriving meaning from complex brain activity data. They offer the enticing prospect of tools that could support and augment human interviewers. In one experiment, for example, researchers demonstrated that ML models could distinguish between truth and lies with reasonable accuracy. The experiment, which was typical of its type, involved a sample of participants who were interviewed about a story they had read. The interviewees were asked either to tell the truth or lie about the story. The interview transcripts were fed into four ML models which analysed different features of the content. The model that was best at detecting liars, with an accuracy of 76 per cent, worked by counting the number of times the interviewees used commonly occurring words.[23] However, the results are perhaps not as exciting as they might seem. Experiments like these involve willing participants who are instructed to lie or tell the truth in low-stakes situations. Those who lie know they are lying as part of an experiment. They also know there will be no adverse consequences if they are found out. They have no skin in the game. The real-life spy who faces a long prison sentence presumably feels different sensations during their interview. AI and ML models are also prone to bias – but then, so too are human interviewers.

Technologies for inferring our thoughts or intentions from our physiological responses or brain activity are still immature. There are good scientific reasons why none of them is yet able to read our minds. In 2022, the UK Information Commissioner's Office (ICO) warned potential users against buying such technologies without first ensuring that they demonstrably work as intended, describing them as 'magic beans', 'hokum', and 'snake oil'. The ICO pointed out that such systems are often biased, inaccurate, and discriminatory, and threatened to take action against organisations that implemented them irresponsibly.[24] That said, AI and ML technologies are evolving in ways that are astonishingly fast and highly non-linear. Who knows where they will be by the time you are reading this?

Self-deception

Another important lesson from science is that the best deceivers are those who deceive themselves. We humans are prone to self-deception. We genuinely come to believe things that are objectively untrue. There are sound biological reasons why evolution has equipped us with the capacity for self-deception, which has been defined as the active misrepresentation of reality to the conscious mind.[25]

Self-deception makes it easier for us to deceive others. The awkward truth is buried in the unconscious mind, while the convenient falsehood is held in the conscious mind and believed by the deceiver. Conscious deception is stressful and hard to conceal, whereas an interviewer cannot easily peer into the unconscious mind of a liar. Self-deception also enables self-promotion. By exaggerating our positive attributes and burying our flaws, it makes us look better to others. Serial liars and charismatic fraudsters make effective use of self-deception, even if they do so unconsciously. Clearly, it is even harder to determine whether someone is lying if they genuinely believe what they are saying. An interviewer faced with a self-deceiving liar is unlikely to uncover their lies without checking their account against independent evidence.

The power of self-deception raises interesting questions about trust. Sensing that a person genuinely believes what they are saying could still be insufficient grounds for trusting them, because they might be deceiving themselves. The safer option is to trust but verify. We might also wonder whether we can always trust ourselves. Our true motivations and intentions might not always be what we think.

Betrayal

Trust is inexorably linked with betrayal. By trusting someone, we implicitly accept the risk of being harmed if our trust is betrayed. We have little choice, because it is impossible to lead a happy and meaningful life, or successfully conduct business, without trusting other people at least a little. In the words of Samuel Johnson, 'it is happier to be sometimes cheated than not to trust'.[26]

Betrayal is a complex and dangerous beast. The concept appears to preoccupy people less now than it did in the days when heresy and treason were words to conjure with. Even so, betrayal remains a universal feature of the human experience.[27] It can range from the minor and quickly forgiven to the devastating and indelible. The severity of a betrayal depends on several variables, including the extent to which the perpetrator consciously intended to betray their victim, the amount of psychological and material harm to the victim, and the depth of prior trust that existed in the relationship. The worst betrayals involve deep deception over long periods, the deliberate and persistent abuse of trusting relationships in pursuit of selfish goals, and the infliction of severe psychological and material harm to the victims. Take Kim Philby, for example.

The Cold War spy Kim Philby betrayed his employer MI6, the UK, and its allies for decades by passing the West's most sensitive secrets to his Soviet handlers. Many western agents died because of him. Perhaps the greatest revulsion for Philby stemmed from his serial betrayal of his family and closest friends. They rallied around him when he came under suspicion, convinced of his innocence, but all along he was playing them for fools.[28] Philby, a charismatic individual who was greatly admired before his treachery was uncovered, was once described as being driven by the 'powerful drug' of deception. In so far as this might be true, he exemplifies one of many reasons why some people become traitors. Betrayal may serve material or ideological interests, but it can also be exciting. The feeling of power and the thrill of the game can be highly seductive.

Shakespeare's works abound with traitors. Othello trusts his seemingly faithful officer Iago ('A man he is of honesty and trust') but Iago fatally betrays him. Duncan trusts Macbeth, but Macbeth fatally betrays him to become king. Shakespeare understood that outward appearances should not be trusted because traitors can conceal their deception even from those closest to them. As Macbeth says: 'False face must hide what the false heart doth know'. John le Carré's 1986 novel *A Perfect Spy* paints a compelling picture of Magnus Pym, a fictional Cold War-era MI6 officer who has secretly been working for the opposition. Pym has spent decades deceiving his wives, his friends, his colleagues, and his employer, all the while concealing his treachery behind layers of inscrutability, lying, misdirection, and iron control over his mood and outward demeanour.

Trusting machines

Automated systems directed by ML or AI are ever-present at work and in our private lives. They perform consequential functions on which our safety or well-being depend. When you obey your smartphone's instruction to turn left, or drive off in a semi-autonomous vehicle, or accept an online insurance quote, you are trusting a machine.

Communication and transport networks, IT systems, smart devices, drones, and robots all require us to trust them, and our lives would become difficult unless we did. Ignoring warnings from intelligent safety systems could be dangerous. However, we could be harmed if these machines made bad decisions, whether through incompetence or hostile intentions. A memorable fictional example is the on-board computer HAL in Arthur C. Clarke's *2001: A Space Odyssey*, which kills to protect its mission. HAL would fit our definition of an insider, as an entity that has been trusted but betrays that trust. Building genuinely trustworthy machines that we can confidently trust is a pressing requirement in an era of ubiquitous AI and lethal autonomous weapons systems. Incidentally, a rule of thumb about current AI is that it tends to be good at doing things that humans find difficult, like quickly analysing huge volumes of data or playing chess, but poor at doing things that humans do effortlessly, like catching a ball or joking with a friend. This is known as Moravec's paradox. It might cease to be true before long.

Does an understanding of insider risk in humans help when deciding how far to trust intelligent machines? The answer is broadly 'yes'. The same basic concepts of trust and trustworthiness that underpin human relationships apply to intelligent artificial systems. Trusting an intelligent machine involves accepting risk by making oneself vulnerable in some way. If we have too little trust in machines, we will fail to benefit from their capabilities. If we trust them too much, we may be harmed if they get things wrong, just as with humans.

We should be willing to trust a machine if we judge it to be trustworthy. A trustworthy machine should have attributes that are equivalent to the trustworthy human's qualities of benign intentions, integrity, competence, and consistency. We should update our estimation of its trustworthiness based on our experience of interacting with it over time. If the machine makes a mistake or acts malevolently, we should trust it less. A truly intelligent machine would then seek to rebuild trust by apologising, explaining its error, and making amends, as would a human.[29] But what do qualities like benign intentions and integrity look like in machines, and how can we assess them?

Computer scientists have been investigating how various characteristics of intelligent machines affect human perceptions of their trustworthiness. The research suggests that our trust in artificial systems, like our trust in other people, is shaped by a mix of subjective beliefs, emotional responses, and cognitive judgements. In other words, it depends on how we feel about the machine and how the machine performs. We are more inclined to trust machines that appear competent, possess human-like features, and have a track record of behaving reliably.[30]

Our emotional responses to machines can lead us astray, just as they do with people. We tend to trust intelligent machines that have anthropomorphic features like faces and voices, even when they perform poorly on objective measures of competence or consistency. Our emotional response to machines that merely *appear* likeable can override our rational judgements. Designers exploit this aspect of our psychology by giving their machines human-like features. However, they are careful to avoid the so-called uncanny valley effect, which is the U-shaped relationship between trustworthiness and human resemblance. We are instinctively uneasy about robots that look very similar to humans but still marginally different. We prefer robots that look more like robots.[31]

Design features that make a machine appear likeable and trustworthy may have no direct bearing on its objective performance, reliability, or safety. Ideally, our judgement would place greater weight on objectively beneficial features, such as safety systems to prevent machines from causing harm or invading privacy, technologies to prevent discrimination or bias, and the ability for independent regulators to audit the machine's behaviour.[32]

Future research on the trustworthiness of intelligent machines should perhaps focus more on the behaviour of the machines themselves, rather than our feelings about them. It is easy to imagine how psychologists, zoologists, anthropologists, and social scientists could study the behaviour of intelligent machines using methods similar to those developed for studying non-human species.[33] The behaviour of robots and other intelligent machines is amenable to observational analysis and experimentation. Some machines can even answer questions.

As technology advances, it seems likely that personnel security will increasingly be called upon to protect organisations against artificial insiders, as well as human insiders. Intelligent machines will vary in their trustworthiness, depending on their functional and design parameters (which could equate roughly with a human's benign intentions and integrity), their competence, and their consistency. Like humans, intelligent machines will sometimes get things wrong and cause harm, for a variety of reasons. All being well, they will learn from their mistakes and improve.

The parallel with human insiders raises the intriguing possibility that artificial insiders will be cultivated and directed by external threat actors, such as hostile foreign states or organised crime groups. External threat actors will be increasingly incentivised to find ways of subverting intelligent machines where once they would have recruited a human insider. They might even use their own intelligent machines to help them do this. Organisations would therefore need to place the equivalent of personnel security defences around their intelligent machines. If so, AI could become the fourth dimension of protective security expertise, alongside physical, cyber, and personnel.

Discussion points

- Is your organisation a high-trust organisation?
- What is the difference between trustworthiness, integrity, and loyalty?
- Who do you trust, and why?
- Does the polygraph work?
- Will we ever be able to read people's minds?
- Can we trust AI?

Notes

1 Luhmann, N. (1979). *Trust and Power*. Chichester: Wiley.
2 Rousseau, D. M. et al. (1998). Not so different after all: A cross-discipline view of trust. *Academy of Management Review*, *23*: 393–404.
3 See, for example: Mayer, R. C., Davis, J. H., and Schoorman, F. D. (1995). An integrative model of organizational trust. *Academy of Management Review*, *20*: 709–734; Colquitt, J. A., Scott, B. A., and LePine, J. A. (2007). Trust, trustworthiness, and trust propensity: A meta-analytic test of their unique relationships with risk taking and job performance. *Journal of Applied Psychology*, *92*: 909–927; Kohn, M. (2008). *Trust. Self-Interest and the Common Good*. Oxford: Oxford University Press; Robbins, B. G. (2016). What is trust? A multidisciplinary review, critique, and synthesis. *Sociology Compass*, *10/10*: 972–986.
4 O'Hara, K. (2004). *Trust. From Socrates to Spin*. Cambridge: Icon Books.
5 Mayer, R. C., Davis, J. H., and Schoorman, F. D. (1995). An integrative model of organizational trust. *Academy of Management Review*, *20*: 709–734.
6 Dietz, G. and Den Hartog, D. N. (2006). Measuring trust inside organisations. *Personnel Review*, *35*: 557–588.
7 Mayer, R. C., Davis, J. H., and Schoorman, F. D. (1995). An integrative model of organizational trust. *Academy of Management Review*, *20*: 709–734.
8 Hillner, L. (2022). Rapport and trust: What's the difference? *CREST Security Review*, *14*: 6–7. crestresearch.ac.uk.
9 Hawley, K. (2014). Trust, distrust and commitment. *Nous*, *48*: 1–20.
10 King-Casas, B. et al. (2005). Getting to know you: Reputation and trust in a two-person economic exchange. *Science*, *308*: 78–83.
11 Colquitt, J. A., Scott, B. A., and LePine, J. A. (2007). Trust, trustworthiness, and trust propensity: A meta-analytic test of their unique relationships with risk taking and job performance. *Journal of Applied Psychology*, *92*: 909–927.
12 Kohn, M. (2008). *Trust. Self-Interest and the Common Good*. Oxford: Oxford University Press.
13 Sinek, S. (2020). www.youtube.com/watch?v=ljLlpOAGRsQ
14 Helliwell, J. F., Huang, H., and Wang, S. (2016). New evidence on trust and well-being. Working Paper 22450. Cambridge MA: National Bureau of Economic Research. www.nber.org/papers/w22450
15 See, for example: Searle, R. et al. (2011). Trust in the employer: The role of high-involvement work practices and procedural justice in European organizations. *International Journal of Human Resource Management*, *22*: 1069–1092; Hope-Hailey, V., Searle, R., and Dietz, G. (2012). *Where Has All the Trust Gone?* London: CIPD; Nerstad, C. G. L. et al. (2017). Perceived mastery climate, felt trust, and knowledge sharing. *Journal of Organizational Behavior*, *39*: 429–447; Kähkönen, T. et al. (2021).

Employee trust repair: A systematic review of 20 years of empirical research and future research directions. *Journal of Business Research*, *130*: 98–109.

16 Funder, A. (2003). *Stasiland*. London: Granta.

17 Vrij, A., Hartwig, M., and Granhag, P. A. (2019). Reading lies: Nonverbal communication and deception. *Annual Review of Psychology*, *70*: 295–317.

18 Meta-analysis is a statistical technique used by scientists. It works by combining the results of many different published studies to produce the strongest possible evidence and conclusions about a particular issue.

19 Bond, C. F. and DePaulo, B. M. (2006). Accuracy of deception judgments. *Personality and Social Psychology Review*, *10*: 214–234.

20 Meissner, C. A. and Kassin, S. M. (2002). "He's guilty!": Investigator bias in judgments of truth and deception. *Law and Human Behavior*, *26*: 469–480.

21 Ford, C. V. (1995). *Lies! Lies!! Lies!!! The Psychology of Deceit*. Washington DC: American Psychiatric Press.

22 FBI. www.fbi.gov/history/famous-cases/aldrich-ames

23 RAND. (2022). *Deception Detection*. RAND Corporation Research Brief. https://doi.org/10.7249/RBA873-1

24 ICO. (2022). 'Immature biometric technologies could be discriminating against people' says ICO in warning to organisations. *ICO*, 26 Oct 2022. https://ico.org.uk/about-the-ico/media-centre/news-and-blogs/2022/10/immature-biometric-technologies-could-be-discriminating-against-people-says-ico-in-warning-to-organisations/

25 Trivers, R. (2013). *Deceit and Self-Deception. Fooling Yourself the Better to Fool Others*. London: Penguin.

26 O'Neill, O. (2002). *A Question of Trust*. Cambridge: Cambridge University Press.

27 Margalit, A. (2017). *On Betrayal*. Cambridge MA: Harvard University Press.

28 Macintyre, B. (2014). *A Spy Among Friends: Kim Philby and the Great Betrayal*. London: Bloomsbury; Milne, T. (2014). *Kim Philby. The Unknown Story of the KGB's Master Spy*. London: Biteback.

29 Kok, B. C. and Soh, H. (2020). Trust in robots: Challenges and opportunities. *Current Robotics Reports*, *1*: 297–309.

30 Glikson, E. and Woolley, A. W. (2020). Human trust in artificial intelligence: Review of empirical research. *Academy of Management Annals*, *14*.

31 Kok, B. C. and Soh, H. (2020). Trust in robots: Challenges and opportunities. *Current Robotics Reports*, *1*: 297–309.

32 See, for example: Toreini, E. et al. (2019). The relationship between trust in AI and trustworthy machine learning technologies. *arXiv:1912.00782v2 [cs.CY]* 3 Dec 2019; Glikson, E. (2022). Emotional overtrust in AI technology. *CREST Security Review*, *14*: 22–23. crestresearch.ac.uk.

33 For an overview of observational methods for studying animal and human behaviour, see: Bateson, M. and Martin, P. (2021). *Measuring Behaviour: An Introductory Guide*. 4th edn. Cambridge: Cambridge University Press.

Part II
Personnel security

6 Personnel security principles

Reader's Guide: This chapter sets out the basic design principles for personnel security. It explains why personnel security should take a systems approach and have a strategic purpose.

Part I of this book discussed the nature and origins of insider risk. Part II considers how organisations can deploy personnel security measures to protect themselves against insider risk. We start in this chapter with a set of general principles that should guide an organisation's approach to personnel security, before progressing in later chapters to the specific measures.

Strategic purpose

Personnel security is the system of protective security measures by which an organisation understands and manages insider risk (as defined in Chapter 1). It forms part of a wider protective security system that also encompasses physical security and cyber security. Personnel security should be an integrated system of complementary capabilities designed to achieve specific outcomes. In practice, however, it often looks more like an assortment of processes and policies that have accumulated over time.

The basic purpose of personnel security is to understand and manage insider risk. To *understand* insider risk, an organisation must be able to gather information about the risk (risk discovery) and make sense of that information (risk assessment). To *manage* the risk, an organisation must be able to act on its understanding by operating defensive measures that reduce the risk to a tolerable level. US authorities have articulated a similar goal for personnel security, which is to 'reduce insider risks to critical assets to acceptable levels'.[1]

In the parallel language of trust, personnel security reduces insider risk by maintaining and building trust. It does this by ensuring that the people who have been trusted with access to an organisation's assets are sufficiently trustworthy, both before and after they are brought on board, and that any potential breaches of trust are detected and prevented at an early stage in their development.

Personnel security regimes that lack any clear purpose or strategy are inclined to underperform because their effort is not directed where it is most needed. Among the least impressive regimes are those with the sole implicit aim of winkling out so-called rotten apples – in other words, the tiny minority of workers who have turned into harmful insiders. The 'rotten apple' meme is questionable in several ways. It falsely implies that insider risk is an inherent property of the individual, ignoring the crucial influence of work and home environments and other external factors. It encourages a binary approach (trusted worker or rotten apple) to a risk that varies along a continuum. Furthermore, it provides ammunition for marketeers selling technologies that purportedly crack the insider problem by locating those rotten apples.

DOI: 10.4324/9781003329022-9

Prevention is better than cure

The best way to manage security risks of any kind is to stop them from materialising, rather than waiting for bad things to happen and dealing with the symptoms. Prevention is generally preferable to cure. The same is true for insider risk. The best personnel security systems are designed to detect the weak early signals of potential insider risk and stop it developing into full-blown insider behaviour. This preventative approach is sometimes referred to as getting 'left of boom'.

A good way of getting 'left of boom' is through a welfare approach, in which the organisation seeks to help people with whatever problems might be nudging them onto the path towards insider action. The welfare approach recognises that most people are never going to become active insiders, and therefore reaching for punitive action at the first sign of trouble is rarely the right answer. On the contrary, a heavy-handed reaction may have the opposite effect if the perceived injustice leaves the individual and their colleagues feeling aggrieved. US authorities have highlighted the dangers of relying on constraints and punishment, because these negative incentives can worsen the risk they are supposed to mitigate. Instead, they advocate a mix of positive and negative incentives, including the fair and respectful treatment of employees.[2] In personnel security, as in life, relying on sticks without carrots is seldom a wise strategy. Parents, teachers, and pet owners can attest that rewarding good behaviour works better, and produces longer-lasting changes, than punishing bad behaviour.

Helping individuals to resolve their personal and work problems before they become an active insider can have a disproportionately positive effect when others in the workforce see their employer treating colleagues humanely. It fosters warmer feelings towards the employer and encourages people to come forward and report any concerns. By means of this so-called bystander effect, the benefits of a welfare approach spread beyond the individuals who are directly helped. Of course, if a person has deliberately flouted the rules and caused significant damage, then a punitive response may be justified, provided it is proportionate and seen to be fair. The point is that punishment should not be the automatic first move.

Deterrence is another way of mitigating insider risk, albeit one whose value is hard to measure or prove. Potential insiders may be less inclined to act if they are fearful of being caught and sanctioned, though deterrence alone will not stop the determined or reckless insider. Organisations can use so-called deterrence communication to strengthen the perception that they take security seriously and have effective measures in place for neutralising threats, including those coming from within. Messages are framed in terms of protecting the well-being of the organisation and its people, with the dual aim of deterring would-be insiders while making everyone else feel that the organisation trusts them and looks after them. Deterrence communication is discussed in Chapter 9.

Despite the obvious attractions of prevention, some organisations have security regimes that mainly deal with the symptoms – for example, by relying on audit and compliance processes to identify those who have broken rules, or using automated monitoring technologies to detect forbidden activities on IT systems. There is nothing inherently wrong with compliance processes or monitoring technologies, although they sometimes promise more than they deliver. The real problem is when they are the only tools in the toolkit. Good personnel security does not involve waiting for an insider risk to materialise and then depending on audit or software to identify the miscreant. Rather, it works strategically to maintain high levels of trust, while also checking that nothing untoward is happening in the shadows.

Believing that prevention is better than cure is not an excuse for neglecting the cure. Prevention is not a substitute for mitigating the vulnerability and impact components of insider risk through measures such as access controls and secure backups. The best protective security strategies aim both to prevent and cure with a system of measures to tackle the threat, vulnerability, and impact components of risk.

Prevention requires action

Detecting the early warning signs of insider risk is of little use unless it triggers appropriate action. Someone must assess the signs and decide what to do in response. However, this vital link between detection and action is often faulty. A recurrent feature of known insider cases, including several outlined in this book, is the presence of clear warning signs that were not acted on until it was too late. For example, an international study of insider fraud found that 85 per cent of fraudsters had displayed at least one behavioural red flag before committing the crime. The commonest red flags for insider fraud were living beyond their means, financial difficulties, and an unusually close association with a vendor or customer.[3]

Failures to act on warning signs often stem from inadequate governance arrangements, where no one feels accountable for the risk or empowered to act. Hindsight is a wonderful thing, and some insiders would always have been difficult to spot, but many cases have slipped through the net despite an abundance of red flags flapping in the wind.[4]

One such example is the MI5 traitor Michael Bettaney, whose case was outlined in Chapter 3. Long before he went through with his betrayal, Bettaney left a trail of clues that should have raised concerns. His colleagues knew him to be a troubled and lonely individual who drank heavily and behaved erratically. He was injured in a bomb attack while serving with MI5 in Northern Ireland in the 1970s, at the height of the Troubles, and might have suffered psychological trauma. Later, while based in London, Bettaney committed a number of minor criminal offences, including public drunkenness and fare dodging. The Security Commission inquiry into the Bettaney affair was critical of MI5's failure to act.[5] The cases of Nidal Hasan and Bruce Ivins, outlined below, are further examples of organisations failing to act on clear warning signs.

Case history

2009: US Army psychiatrist Major Nidal Hasan shot and killed 13 of his army colleagues and wounded 43 others at Fort Hood, Texas, in the worst terrorist attack on US soil since the 9/11 attacks of 2001. His path to violent insider action was strewn with warning signs. Over the years preceding the attack, Hasan was vocal in his support for radical Islam and his admiration of Osama bin Laden. He stated in front of army colleagues that his commitment to Sharia law overrode his commitment to the US Constitution and argued that suicide bombings were justifiable. One colleague described him as a ticking time bomb; another called him a religious fanatic. Several reported their concerns, but nothing was done. A year before Hasan's attack, the security authorities discovered that he had been in sustained email contact with Anwar al-Awlaqi, a leading figure in Al Qaeda based in Yemen. In one email, Hasan asked whether a Muslim US Army soldier would be considered a martyr if he committed fratricide. No one interviewed Hasan about the emails, which were dismissed as innocent research. To cap it all, Hasan was known to be a chronically poor performer at his job, who often turned up late. Despite this, he continued to receive positive appraisals and was promoted a few months before the attack.[6]

Comments

- The US Army and other federal agencies ignored a series of clear warning signs that should have prompted decisive action.
- Subsequent investigations highlighted failures of leadership, governance, policy, and culture as root causes of the collective failure to act. Everyone appeared to think it was someone else's responsibility.

Case history

2001: Soon after the 9/11 terrorist attacks on the US, letters containing anthrax spores were sent through the mail to several newspaper offices and US senators, causing widespread contamination. Five people who came into contact with the letters died of anthrax and at least 17 others were infected. The cost of decontaminating numerous buildings was enormous. Several years later, after a huge investigation, the US authorities identified the perpetrator as Dr Bruce Ivins, a scientist working for the US Army Medical Research Institute of Infectious Diseases. Ivins was a microbiologist who had spent years researching anthrax. He killed himself in 2008, a few days before he was due to be charged. It later emerged that there had been numerous red flags indicating that Ivins was a risk. He was an eccentric loner with a history of heavy drinking and significant mental health problems. He received treatment from several psychiatrists and therapists, whom he told of his recurring thoughts about killing people and his plans for acquiring weapons. He committed criminal offences, including breaking and entering, theft, and vandalism. He became obsessed with women whom he then stalked. In the months before the attack, Ivins unusually spent long periods alone in the secure biocontainment labs, often at night or weekends. He killed himself days after being released from a psychiatric hospital, to which he had been involuntarily committed as a danger to himself and others. The anthrax letters have been described as an example of a 'predictable surprise': an event that was heavily signposted in advance but still caught people unaware. Ivins's motivation for the attack remains a mystery, although some evidence suggests he intended it to be a wake-up call to the authorities rather than a mass-casualty attack. Given his expertise with anthrax, one of the deadliest biological weapons, he presumably could have killed far more than five people if that had been his intention.[7]

Comments

- No decisive action was taken despite many warning signs.
- The repeated failures to act appeared to stem from organisational and cultural problems, including excessive reliance on inadequate vetting procedures, failure to share information between investigators and Ivins's therapists, over-familiarity with Ivins as a longstanding colleague, and a culture of complacency about insider risk.
- Even without the benefit of hindsight, it is remarkable that someone with Ivins's track record was granted the security clearances needed to work on a weapon of mass destruction in a sensitive military laboratory.
- He might not have intended to kill people. If so, this would be an example of an insider attack that had unintended consequences.

Prevention or reconciliation?

What if prevention does not work, and our putative insider stays on the developmental path to insider action? Are there other ways of mitigating the risk, short of waiting to detect the insider's hostile actions? Prevention is never guaranteed, and we know that capable insiders can escape detection for years, or altogether. Most spies have been discovered not through detection, but from intelligence provided by defectors or penetration of the adversary. It would therefore pay to explore other options.

One interesting idea that has been advocated by the American psychiatrist David Charney is to offer insiders – specifically insider spies involved in state espionage – the prospect of an escape route, or off-ramp. The concept, which Charney calls reconciliation, is to establish a formal process by which spies could hand themselves over to the authorities, knowing they would face penalties less severe than if they were caught later. Reconciliation is intended to reduce the security risk by lowering the barriers to exit before the insider causes more severe damage.[8] Sceptics might argue that offering insiders a safer alternative to being caught and severely punished would weaken deterrence. In some cases, however, it might still be preferable to waiting until serious harm has been done, especially when the legal or disciplinary options are uncertain.

The age-old practice of 'turning' spies into double agents and directing them back against their former spymasters could be viewed as a form of reconciliation – at least, in those cases where the spy volunteers to the people they were meant to be spying on. During World War Two, MI5 had great success with turning captured German agents and using them to penetrate and mislead German intelligence. One case, which later inspired a Hollywood movie, was that of Eddie Chapman. Early in the war, Chapman, a career criminal who was wanted by the British police, volunteered to spy for the Germans. He was taken abroad by the Germans, trained in spying and sabotage, and parachuted back into England, where he immediately handed himself in to the authorities. He volunteered to work for MI5, who sent him back to Germany, masquerading as a loyal agent of German intelligence. Chapman then spied on the Germans, who still believed he was working for them. They even awarded him their highest honour, the Iron Cross. Chapman was known to MI5 by the codename Zigzag, which was fitting in view of his switches in allegiance.[9]

Holistic, dynamic, and adaptive

Holistic security

A central doctrine of protective security states that physical, personnel, and cyber security risks should be managed holistically rather than separately, recognising their huge interdependencies. Security professionals subscribe to this doctrine because it makes sense and they know from experience that it works. A well-placed insider can defeat physical or cyber security defences; cyber attacks can facilitate physical or insider attacks; physical and personnel security measures are needed to protect cyber systems; and so on. Personnel security should form part of a holistic protective security system in which the physical, cyber, and personnel domains are managed collectively rather than in separate silos.

Case histories

2022: When Russia invaded Ukraine in February 2022, the military attack was accompanied by cyber attacks on Ukrainian government organisations, aimed both at stealing military secrets and causing disruption. Experts believe that the Russian cyber attacks were augmented and facilitated by insiders who had infiltrated Ukrainian ministries.[10]

2010: It emerged that a prolonged and sophisticated cyber attack on secret Iranian facilities had disrupted the Iranian nuclear weapons programme. The attack, which became known as Stuxnet, worked by interfering with electronic systems controlling centrifuges used to refine uranium for nuclear bombs. The systems were isolated from the Internet and hidden within a mountain inside Iran. It is thought that the malware used in the attack found its way onto the Iranian systems through witting or unwitting insiders.[11]

Comments

- External threat actors may deploy blended attack methods in which insiders facilitate cyber attacks.
- Insiders can circumvent cyber security and physical security.

Even though the case for holistic security is compelling, many organisations have security structures that are far from holistic, with cyber security sitting in one silo and physical security in another, while the personnel security Cinderella languishes homeless or straddles several silos. The gulf between the doctrine of holistic security and the practice of separate stovepipes is illustrated in Figure 6.1, which also depicts the general ascendancy of cyber security over physical and personnel security. The fragmentation of protective security into separate organisational silos is a recipe for deficient security.

Dynamic and adaptive security

In common with other types of security risk, insider risk is dynamic and adaptive. As noted in Chapter 1, the risk changes over time and it adapts in response to the defensive actions of the potential victims. Intentional insiders are intelligent threat actors who try to avoid detection and find ways of defeating security. In some cases, their ability to do this is enhanced by support from a sophisticated external threat actor. For personnel security to have any hope of gripping insider risk, it too must be dynamic and adaptive. That requires, among other things, agile mechanisms for discovering and assessing risks, making decisions, and learning from experience. These capabilities are described in Chapter 9.

The systems approach

It might be tempting to believe that a single piece of technology, such as automated monitoring software, or a single process, such as pre-employment screening, can deal with insider risk. Tempting but wrong. Both in practice and in principle, no single process or technology by itself can ever provide an adequate defence against insider risk. Personnel security requires defence in depth from a system of complementary measures and supporting capabilities. The fundamental reason is that insider risk – in common with most non-trivial problems – is an emergent property of a complex system. It is worth pausing to examine what this means.

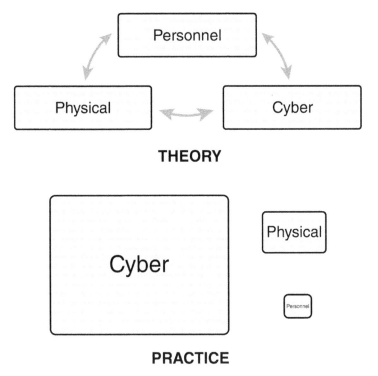

Figure 6.1 Theory (holistic security) versus practice (separate silos)

Complex systems

A complex system consists of interacting components whose collective behaviour is more than the sum of its parts. Complex systems have emergent properties that cannot be deduced from the characteristics of their individual components alone. For instance, consciousness is an emergent property of the human brain that cannot be inferred from the properties of individual neurons. Complexity should not be confused with mere complication. A mechanical clock is *complicated*: it can be designed and built, and it will behave in predictable ways. However, brains, businesses, terrorist networks, the weather, organised crime groups, economies, governments, and hostile foreign states are *complex*. They have higher-order properties that are not shared by merely complicated systems like clocks and engines. The most relevant of these properties are emergent behaviours and non-linear change.

Under some conditions (known mathematically as chaos), complex systems become highly sensitive to small disturbances, making it impossible, both in principle and in practice, to predict their longer-term behaviour in detail.[12] This helps to explain why economists have such a poor track record of predicting the behaviour of national economies, and why meteorologists cannot provide a detailed local weather forecast for a year ahead (although they can predict the overall climate). The extreme sensitivity of chaotic complex systems is sometimes referred to as the butterfly effect. The name comes from the title of a 1972 paper by the mathematician Edward Lorenz called 'Does the flap of a butterfly's wings in Brazil set off a tornado in Texas?' Using computer models, Lorenz showed that even tiny changes in starting conditions could produce huge non-linear and unpredictable changes in the subsequent behaviour of weather systems.

Some complex systems also possess an ability to adapt through learning or evolution. They are known as complex adaptive systems. A tropical storm is a complex system, but a hostile foreign state is a complex *adaptive* system. In complex adaptive systems, including those made of interacting humans, the individual components respond and adapt to each other. Complex adaptive systems cannot be designed and controlled in the same way as complicated machines, and their behaviour may be unpredictable.

Policymakers sometimes appear to ignore the nature of complex adaptive systems. One example is the illusion of control, whereby policymakers and leaders persist in believing they can control the behaviour of complex adaptive systems by pulling a single policy lever and expecting a specified outcome. The real world does not behave like that. Well-intentioned efforts to solve complex economic or social problems sometimes have unintended consequences that make the problem worse.

Insider risk involves interactions between people acting within an organisation and their wider environment. As such, it is a phenomenon that emerges within a complex adaptive system. Managing that risk requires a systems approach involving interlocking and adaptive capabilities, including risk discovery, risk assessment, governance, and leadership. No single policy, process, or technology could be sufficient by itself. To put it another way, there are no silver bullets in personnel security. Beware of anyone who claims there are.

The systems approach to personnel security is a crucial concept and worth dwelling on a little longer. An analogy may help.

The immune system analogy

The immune system is a fabulously complex adaptive system that has evolved to protect humans and other animals from viruses, bacteria, fungi, parasites, cancer cells, and other potentially harmful invaders. It must reliably protect us against a vastly diverse and constantly evolving range of threats. It does this by means of multiple, complementary layers of defence. The outer layers include so-called innate immune cells, such as macrophages, which can trigger inflammation and engulf foreign material. They provide a rapid first line of defence against a wide range of threats. The inner layers are adaptive: they mount highly tailored responses to novel and specific threats, like new strains of viruses or bacteria. The adaptive immune system involves many different types of specialised cells that work together in highly coordinated ways. They include B-cells, which form antibodies against specific pathogens; T-cells, which attack infected and cancerous cells; helper cells, which coordinate the actions of other immune cells; and regulatory T-cells, which damp down the immune response once the threat has passed. To add to the complexity, the immune system and the brain are interconnected and influence each other in countless ways.[13]

Like security, the immune system's response to a threat must be correctly calibrated. If it under-reacts, a dangerous infection might take hold, but if it *over*-reacts then the immune system might attack its own body, giving rise to an allergic reaction, auto-immune disease, or chronic inflammation. In the analogous world of personnel security, under-reaction is more common than over-reaction. Even so, problems can arise if an organisation's response to a perceived threat is excessive, or if it sees threats where none exist. An organisation in which innocent people are subject to unfounded suspicions, and no one trusts anyone else, can tear itself apart, as the CIA discovered in the 1960s when the agency became convinced it had been heavily penetrated by Soviet spies. A healthy balance must be struck between naïve over-trusting and paranoid suspicion. The judgements behind that balancing act should rest on solid evidence and careful assessment.

The immune system learns from experience. When it encounters a new bacterial or viral threat, it responds adaptively and remembers how to deal with that threat if it should encounter it again, possibly years later. The immune system's response is optimised through frequent exposure to a wide variety of threats that are below the level needed to cause serious disease. One further analogy with protective security (you can no doubt think of more) is that we only notice it is there on the rare occasions when it fails to protect us.

Personnel security systems

What does a personnel security system look like in practice? Various models have been proposed which set out the key components. One of the best-known of these is the Personnel Security Maturity Model (PSMM) produced by the National Protective Security Authority (NPSA, formerly CPNI), the UK government's national technical authority for personnel and physical protective security. The PSMM consists of seven 'pillars', each representing a critical component of personnel security.[14] The seven pillars are:

- Governance and leadership
- Insider risk assessment
- Pre-employment screening
- Ongoing personnel security
- Monitoring and assessment of employees
- Investigation and disciplinary practices
- Security culture and behaviour change

The PSMM covers the essential building blocks of personnel security. No organisation could claim to have comprehensive protection without possessing at least some capability under all seven headings.

The same set of capabilities can be looked at in a different way, through the lens of trust. An even simpler model divides the components of a personnel security system into three broad categories:

- *Pre-trust measures*: protective security measures applied *before* an organisation trusts a person by giving them access. These are sometimes referred to as pre-employment screening or due diligence.
- *In-trust measures*: protective security measures applied *after* a person has been trusted and given access. These are sometimes referred to as aftercare or ongoing security.
- *Foundations*: cross-cutting capabilities that underpin the whole system, including risk management, governance, leadership, and culture.

The simple threefold model of personnel security is represented in Figure 6.2.

Understanding the benefits

Personnel security costs money and imposes frictional overheads. Organisational leaders naturally want to know how its benefits stack up against the more obvious costs. Some organisations fail to invest sufficiently in personnel security because they have not understood its benefits.

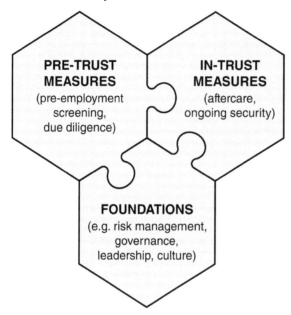

Figure 6.2 A simple threefold model of a personnel security system

In broad terms, the three main reasons for investing in personnel security are:

1. Most obviously, personnel security mitigates insider risk and therefore reduces the harm that insiders would otherwise be likely to cause. This is the traditional case for protective security – that it stops bad things from happening.
2. Good personnel security helps to maintain and build trust, bringing wider business benefits beyond the reduced security risk. As we saw in Chapter 5, high-trust organisations tend to be more innovative, faster at making decisions, less bureaucratic, and better able to recruit and retain the best people.
3. Personnel security strengthens organisational resilience (see below).

Advocates of personnel security point out that it is cheap in comparison to physical and cyber security. (A cynic might wonder if it is neglected *because of* its cheapness.) Good personnel security does not require lots of expensive infrastructure, although the right technology can certainly help. It is primarily about changing people's attitudes and behaviour. Leaders should bear in mind that the damage resulting from a major insider attack could far exceed the cost of building a half-decent personnel security system.

Strengthening resilience

Personnel security helps to strengthen organisational resilience. What does this mean? Resilience is another of those slippery words for which there is no universally agreed definition. One way of thinking about resilience is the organisation's ability to continue delivering its mission despite stress and disruption. Another way is to regard resilience as the capability to avoid crises and cope better with the crises that cannot be avoided. (Again, prevention is better than cure.)

A traditional view is that resilience is about recovering from disruption and returning to normality. This passive form of resilience derives from capabilities like business continuity planning, incident management, crisis management, and disaster recovery. A more interesting form of resilience, known as active resilience, goes beyond passive resilience by adding the capability to grow progressively tougher by learning from adversity and becoming better able to cope with future stresses.[15] Active resilience is about being adaptive. Personnel security strengthens resilience by helping the organisation to avoid crises arising from damaging insider attacks, and cope better with the crises they cannot avoid; for example, by reducing the impact of insider attacks and recovering rapidly from them.

Discussion points

- Why do most organisations focus more attention on cyber security than on insider risk?
- Do conventional risk management processes take sufficient account of the adaptive nature of insider risk?
- What are the key features of complex adaptive systems and what lessons can we learn for security?
- What percentage of your organisation's budget goes on personnel security?
- Is your organisation's security holistic?

Notes

1 Costa, D. (2020). *Applying the CERT Resilience Management Model to the Counter-Insider Threat Mission*. Carnegie Mellon University. https://apps.dtic.mil/sti/pdfs/AD1110237.pdf
2 Moore, A. P. et al. (2018). Balancing organizational incentives to counter insider threat. *2018 IEEE Security and Privacy Workshops (SPW)*. San Francisco CA. https://doi.org/10.1109/SPW.2018.00039.
3 ACFE. (2022). *Occupational Fraud 2022: A Report to the Nations*. https://legacy.acfe.com/report-to-the-nations/2022/
4 Bunn, M. and Sagan, S. D. (2016). A worst practices guide to insider threats. In *Insider Threats*, ed. by M. Bunn and S. D. Sagan. Ithaca NY: Cornell University Press.
5 Security Commission. (1985). *Report of the Security Commission*, May 1985. Command Paper 9514. London: HMSO.
6 Zegart, A. B. (2016). The Fort Hood terrorist attack. An organizational postmortem of Army and FBI deficiencies. In *Insider Threats*, ed. by M. Bunn and S. D. Sagan. Ithaca NY: Cornell University Press.
7 Stern, J. and Schouten, R. (2016). Lessons from the Anthrax Letters. In *Insider Threats*, ed. by M. Bunn and S. D. Sagan. Ithaca NY: Cornell University Press.
8 Charney, D. L. (2019). *Noir White Papers on Insider Threat, Counterintelligence and Counterespionage*. www.NOIR4USA.org
9 Macintyre, B. (2007). *Agent Zigzag*. London: Bloomsbury.
10 Grylls, G. (2022). British cyberwarriors defend Ukraine. *The Times*, 1 November 2022.
11 Corera, G. (2016). *Cyberspies: The Secret History of Surveillance, Hacking, and Digital Espionage*. NY: Pegasus.
12 Mitchell, M. (2009). *Complexity: A Guided Tour*. Oxford: Oxford University Press.
13 Martin, P. (2005). *The Sickening Mind: Brain, Behaviour, Immunity and Disease*. London: Harper Perennial.
14 NPSA. (2023). www.npsa.gov.uk/personnel-security-maturity-model
15 Martin, P. (2019). *The Rules of Security: Staying Safe in a Risky World*. Oxford: Oxford University Press.

7 Pre-trust measures

Reader's Guide: This chapter describes the pre-trust personnel security measures (also known as pre-employment screening) that can be applied before a person is trusted with access to an organisation's assets.

The simple threefold model outlined in Chapter 6 divides the panoply of personnel security defences into three categories: pre-trust measures, in-trust measures, and foundations. This chapter looks more closely at the first of these.

Pre-trust measures are the protective security measures applied *before* an organisation trusts someone by giving them access to its assets. The main types of pre-trust measures are:

- Interviews
- Record checking
- Open-source intelligence
- Psychometric testing

The terms pre-employment screening, vetting, and due diligence may be roughly equivalent in meaning to pre-trust measures, although in some contexts they refer only to the record-checking element. Vetting is an ambiguous and rather stale term, which can mean anything from record checking to personnel security in its fullest sense. It is best to be explicit.

Pre-trust risk indicators

The purpose of the pre-trust measures described in this chapter is to deter and detect potential insiders before they are trusted with access. Deterrence aims to do this by discouraging would-be insiders from applying in the first place. The presence of pre-employment screening might deter those who are sufficiently concerned about incriminating aspects of their past being revealed, but it should not be relied on. Deterrence can be strengthened with deterrence communication, which is described in Chapter 9.

If deterrence fails, and a potential insider enters the recruitment process, then pre-trust measures are there to detect signs that the individual might present a heightened security risk. The types of risk indicators that pre-trust measures are intended to detect include:

- Dishonesty (e.g. lying about identity, qualifications, work experience, illicit drug use, or criminal convictions)
- Significant criminality (e.g. convictions for serious crimes of dishonesty or violence)

DOI: 10.4324/9781003329022-10

- Association with threat actors (e.g. connections with organised crime, terrorism, extremism, or a hostile foreign state)
- Financial vulnerability (e.g. large and uncontrolled debt that might make them vulnerable to pressure)
- Psychological vulnerability (e.g. personality characteristics, mental health problems, or substance abuse issues that might predispose them to insider action)
- Suspicious travel (e.g. undeclared travel to places where they might have been exposed to hostile foreign state actors)

Pre-trust measures by themselves can never provide a robust defence against insider risk. As we saw previously, most insiders become insiders after they join an organisation, and even the best-intentioned people can be tricked or coerced into becoming insiders. Furthermore, pre-trust measures could be subverted by an existing insider who exploits their authority to recruit an individual who would not otherwise satisfy the organisation's entry requirements. The main work of personnel security therefore starts once the person is on board. Nonetheless, every organisation should conduct at least basic due diligence, if only to deter potential infiltrators and avoid the embarrassment of unwittingly employing a serious criminal or terrorist.

Interviews

Most professionally run organisations would interview a person before giving them a job. However, they might not always interview a potential contractor or supplier in the same way, even though these roles may afford significant access. Either way, conventional recruitment interviews are not designed to assess a candidate's trustworthiness or potential insider risk. Rather, they are intended to judge the candidate's skills, suitability for the role, and 'fit' with the organisation. A potential insider might tick the right boxes for professional competencies and interpersonal skills yet still be untrustworthy. Very little security comfort should therefore be drawn from a standard recruitment interview, which is an exceedingly weak tool for judging insider risk.

Unstructured interviews

An unstructured interview that amounts to little more than a chat is a deceptively poor way of assessing a person's likely future work performance, let alone their integrity. In scientific terms, it has low predictive validity as a measurement tool. It is unlikely to offer more than a subjective impression, which may be wrong.[1] One reason is the halo effect – our predisposition to place excessive faith in our initial impressions. The halo effect is well known, but that does not stop us from succumbing to it. Another problem is confirmation bias – our predisposition to disregard information that conflicts with our existing beliefs. The interviewer reaches an implicit decision early in the interview, based on initial impressions, and subsequently disregards information that might cause them to change their mind. Psychological predispositions and cognitive biases are discussed in Chapter 11.

   To give some idea of just how subjective our judgements can be, consider this. A carefully conducted scientific experiment found that when people reviewed CVs presented to them on a heavy clipboard, they rated the candidates more favourably, and were more confident in their judgements, than when the CVs were presented on a *light* clipboard.[2] Other studies have shown that our judgements are affected by such extraneous factors as the time of day and room temperature. Despite their many weaknesses, unstructured interviews are convenient, and they remain in widespread use.

Structured interviews

What about a more structured approach? The power of an interview can be substantially improved by focusing on collecting information, rather than forming subjective impressions that lead to an immediate decision. However, the information collected in a conventional recruitment interview is intended to inform judgements about skills and competencies, not trustworthiness. A prevalent HR doctrine decrees that, in the interests of fairness and transparency, recruitment interviews should be structured around so-called core competencies, with all candidates being asked essentially the same set of questions about areas of ability like delivery, teamwork, managing change, and communication. Probing the awkward details of an individual's beliefs or ethical standards is discouraged. Competency-based recruitment interviews are virtually useless for judging trustworthiness, except when they reveal that a candidate has invented evidence of their competence.

If an interview is to be used for judging trustworthiness, then it should be designed and conducted with that purpose in mind. High-security organisations like intelligence agencies conduct additional interviews specifically to assess the candidate's trustworthiness and search for evidence of security risks. Some conventional organisations do the same when recruiting for high-risk positions that give access to their crown jewels. But even a security-focused interview can provide only limited assurance by itself, because it is surprisingly difficult to tell whether someone is honest just by talking to them.

As we saw in Chapter 5, our ability to detect deception is weaker than we think. Most people, including security professionals, perform poorly when determining whether a person is lying in a conventional interview. Their performance is even worse if they believe in the significance of non-verbal cues like gaze aversion, body posture, and fidgeting. These supposedly tell-tale signs are in fact weak and unreliable indicators of deceit.[3] A high-stakes interview can make even the most honest person feel anxious, and their non-verbal behaviour may reflect this.

Fortunately, there are evidence-based techniques that can improve the interviewer's prospects of determining whether a person is telling the truth. The techniques that work best start from the premise that non-verbal behaviours are largely irrelevant, and the interviewer should instead focus on the content of what the interviewee says. These techniques are designed to amplify and expose the different psychological states of liars and truth-tellers, especially their willingness to provide detailed and verifiable information. In essence, liars want to conceal information that might expose them as liars, whereas truth-tellers want to demonstrate their honesty. Hence, liars tend to keep their stories shorter and lighter on detail. They do this both to make their stories easier to recall and to avoid exposing any loose ends that might reveal their dishonesty.[4]

One technique that can assist in distinguishing truth-tellers from liars is to impose a high cognitive load on the interviewee – in other words, place heavy demands on their thinking capacity.[5] The aim is to flush out deception by exploiting limitations on our capacity to recall and think about several things at the same time. This is done with tactics like requesting lots of background detail, posing quirky and unexpected questions, or asking for an account to be repeated in a different chronological order. Detailed lies are harder to concoct and harder to recall accurately than the truth, and even more so when having to think simultaneously about other things. As the sixteenth-century philosopher Michel de Montaigne observed: 'you can normally hobble liars by making them tell the same tale several times over ... he who does not feel his memory to be strong enough has no business lying'.

Another technique, known as Strategic Use of Evidence, involves comparing what the interviewee says against independent factual evidence compiled before the interview, without revealing what is already known. Liars are wary of exposing discrepancies in their account and are therefore reluctant to offer up details. The interviewer asks questions to which they already know the answers and listens for discrepancies in the response. A related technique known as the Verifiability Approach is designed to flush out the liar's reluctance to provide details that can be verified, like the names of other people.[6] It is based on evidence that the accounts of liars typically contain fewer verifiable details than those of truth-tellers, both in absolute terms and as a proportion of all details given. Liars tend to give vague, sketchy accounts whereas truth-tellers paint a richer picture. Two other techniques, known as Cognitive Credibility Assessment and Reality Interviewing, are designed to probe the quality and richness of the interviewee's statements and expose the liar's tendency to keep their stories short on details, especially details about sensory experiences like sounds, smells, or tastes.[7]

To summarise: one way of assessing a person's trustworthiness is to find out if they tell the truth; and one way of doing that is to conduct a structured interview along the following lines:

- Listen. Focus on the content of what they say, not how they say it or how they behave. Non-verbal behaviour is a weak guide to deception.
- Compare what they say against independently verified evidence, without revealing in advance what you already know.
- Ask for verifiable details. Liars try to avoid providing loose threads for the interviewer to pull on.
- Ask them to describe sensations and context: colours, smells, tastes, spatial relationships, chronological sequences, and so on. Liars find these harder to invent and recall.
- Encourage them to say more. Truth-tellers are more forthcoming; they want to demonstrate their honesty, although they do not necessarily give the complete story the first time around.
- Ask for the same information more than once and in different ways. Listen for discrepancies or inconsistencies.
- Ask unexpected questions. Liars find these harder to deal with, whereas truth-tellers answer expected and unexpected questions equally well.
- Put them under cognitive pressure; for example, by asking quick-fire questions or getting them to describe events in reverse order. Liars find it harder to sustain lies under high cognitive load.
- Consider recording the interview or including a second interviewer who only listens. A listener is less distracted by irrelevant non-verbal behaviours and consumes less of their cognitive capacity in direct interactions with the interviewee.

Recruitment: meet security

The recruitment process in many organisations maintains a bureaucratic separation between the responsibilities of HR and those of security. Conventional recruitment is about selecting people with the right skills and competencies for the job. Pre-trust personnel security measures, such as national security vetting or due diligence checks, are often managed separately. The implicit assumption behind this division of roles is that judgements about suitability and security are separate and distinct. This assumption is questionable, however, because some personal characteristics are clearly relevant to both. For instance, it would be unwise to hire someone who lacks integrity and does not share the ethical values of the organisation. The mismatch between their character and the culture of the organisation could adversely affect their performance and work relationships, while their lack of integrity could make them a security risk. Either way, they would not be a promising candidate.

In view of the overlap, there is an argument for bringing together the HR-led and security-led elements within an integrated function for selecting and recruiting the right people – that is, people who are both competent and trustworthy. A further advantage would be better sharing of relevant information and less transactional friction between organisational silos. However, better integration of the HR recruitment and personnel security functions does not mean merging them into an amorphous blob or handing everything over to HR. The two functions require different attitudes and skills, and there is an inherent tension between HR's drive to fill vacancies and security's need to exclude untrustworthy candidates. They are not the same, but they are definitely complementary.

Record checking

The second main category of pre-trust measures is record checking. It is the workhorse of pre-trust measures. As its name suggests, record checking involves checking records and documents for relevant information about the individual. This is done for two main reasons: first, to verify the person's identity and legal right to work; and second, to uncover any evidence of past behaviour that would indicate a heightened security risk.

Verifying identity and right to work

Most organisations take the basic precaution of confirming that an applicant is who they say they are, and that they are legally permitted to work. This is normally done by inspecting such things as their passport, driving licence, birth certificate, and any official immigration documents. A small minority of people lie about their identity or right to work, and present forged documentation to support their claims. Differentiating between genuine and falsified documents is not always easy and may require training. Obviously, someone who lies about their identity is unlikely to be trustworthy.

Evidence of security risk

A more direct way of assessing trustworthiness is by checking official records for evidence of past behaviour suggesting the individual presents a security risk. In the absence of other reliable evidence, the best guide to future behaviour is past behaviour – in particular, behaviour that signifies dishonesty, lack of integrity, criminality, or close association with threat actors.

A basic precaution is to verify the individual's account of their educational and employment history. A surprising number of recruitment candidates exaggerate or lie about their past achievements and make false claims about their qualifications. The hiring organisation should exercise due diligence by asking referees to confirm employment history and by checking documents like degree certificates and professional diplomas. Again, if a person is found to have lied or exaggerated then they are unlikely to be trustworthy.

Pre-employment screening of this sort should be a given – the minimum precaution that any responsible organisation would take. But some organisations do not even do the minimum. For example, an international study of insider fraud found that 43 per cent of organisations that had been victims of insider fraud had not run a background check on the perpetrator before hiring them. One in five of the fraudsters had previous red flags that should have been picked up.[8] Some UK police forces were criticised in a 2022 official report for not routinely interviewing their recruitment candidates, despite multiple insider scandals.[9]

Case histories

2023: A psychiatrist working in the UK National Health Service was jailed for fraudulently obtaining employment by forging her medical qualifications and documentation. She had worked as an NHS psychiatrist for more than 20 years, despite lacking the necessary medical qualifications, and was paid more than £1M during this time. Before her deception was finally uncovered, she was the subject of complaints and investigations, none of which resulted in decisive sanctions.[10]

2022: An airline pilot with British Airways subsidiary BA CityFlyer was jailed for fraudulently overstating his previous flying experience to obtain his job. He made false entries in his logbook to make it appear that he had more flying hours.[11]

2021: British police officer Benjamin Monk was jailed for manslaughter. It emerged that he had failed to disclose his previous criminal record when he joined the police.[12]

2021: Metropolitan Police officer Ben Hannam was jailed for membership of a banned neo-Nazi group. His connection with the extreme right predated his recruitment to the force. He appeared in a neo-Nazi propaganda video days before applying to join the police in 2017.[13]

Comments

- Inadequate pre-employment record checking may enable unqualified or unsuitable people to gain entry, causing reputational damage to their employer and potential harm to the public.
- Once on board, fraudulent recruits may become progressively harder to detect as they build up credibility.

A more rigorous form of record checking involves inspecting criminal records to see if the individual has been convicted of crimes they have not declared in their application. Many employers routinely ask job applicants to declare any significant criminal convictions and then check criminal records to see if the applicant has concealed any. In the UK and other countries, there are legal restrictions on what types of offences an employer can ask an applicant to declare and what types of criminal records they can inspect. In the UK, the process normally excludes so-called spent convictions, which means criminal convictions that are deemed to have expired with rehabilitation and a legally determined passage of time. Spent convictions do not show up on a basic criminal record check. However, a conviction *will* show up on a basic check if it is for a serious offence or one that has not yet expired. Convictions for violent or serious sexual crimes, or offences carrying a long prison sentence, stay on the individual's record indefinitely.

Having some form of criminal record does not automatically make someone an unacceptable security risk. It depends on the nature, frequency, and recency of the offence or offences. Individuals with historical convictions for, say, minor road traffic or drug offences can still be granted the highest level of government national security vetting clearance. Employers must be pragmatic, given that a surprisingly large proportion of the adult population has a criminal record. In England and Wales, for example, one in three men born in the 1950s had a criminal record by the time they were middle-aged.[14]

A different type of scrutiny comes from inspecting the individual's financial records, which they are asked to supply. The aim is to identify potential risk factors like large and unsustainable debts or suspicious payments. Traditional vetting procedures have tended to place considerable weight on personal finances. However, the rationale for doing so may be uncertain. The causal connection between financial difficulties and insider action is not always clear. As we have seen, money is not the main driver in many insider cases and most people with financial problems do not become active insiders. Moreover, the goal posts have moved over time: levels of personal debt that were once regarded as worryingly high have since become normal, as more people enter employment with big student debts and take on barely affordable mortgages. The size of a debt matters less than the individual's ability to manage it. Perhaps one reason why personnel security officials have been drawn to financial records is that they offer a safe space of objective evidence. It may feel more comfortable probing the size of a person's overdraft than their attitudes and beliefs.

Government organisations are permitted to dig deeper with record checking in the interests of national security. Applicants for roles with access to highly classified information are subject to a national security vetting process that additionally includes checks against intelligence records for terrorism, hostile foreign state activity, and other national security threats. Law enforcement agencies check their criminal intelligence records for links with serious and organised crime.

Limitations of record checking

Even the most thorough record checking has two limitations: it is only as good as the records that are checked; and it is only valid on the day the check is made. An absence of evidence is not evidence of absence, so a failure to find adverse records does not prove the person is trustworthy. It may just be that they were never caught. Moreover, a conventional record check is made at a point in time and presents only a snapshot. An individual might go on to behave badly even though there was no record of wrongdoing at the time of checking. An aspiration for future vetting is to make the record checking continuous, so that new misdemeanours show up as soon as they are recorded.

Despite these weaknesses, many organisations place excessive reliance on record checking, as though an absence of adverse historical records at a point in time somehow constituted proof of the individual's current and future trustworthiness. It does not.

Government and police vetting

Most governments seek to protect national security with personnel security arrangements, which they often refer to as vetting. The UK has had a formal national security vetting process since the mid-twentieth century, and probably much longer in a less formal guise.[15]

The procedures for national security vetting rely heavily, though not exclusively, on record checking. In the UK, the lowest level of national security vetting involves checking the individual's identity, right to work, employment history, criminal records, and national security records. The next level up adds financial checks. Only at the highest level of vetting clearance, required for those with frequent and uncontrolled access to the nation's most sensitive secrets, does the process include security interviews with the individual and referees.

National security vetting in the UK has a longstanding reputation for being slow and cumbersome. Clearances can take weeks or months, with further delays if a record check uncovers possible adverse traces that must be assessed by an expert. A 2023 report by the UK National Audit Office (NAO) on the performance of UK Security Vetting (UKSV), the organisation that carries out national security vetting checks for UK government departments and other public bodies, found that most clearances and renewals took far longer to process than their target times. For example, it took an average of eight months to renew a top-level Developed Vetting (DV) clearance. The long delays made it difficult for departments to fill vacancies, with consequential effects on the delivery of public services. The NAO blamed the poor performance in part on UKSV's reliance on an old and unreliable IT system. Previous attempts to replace the IT and 'transform' the vetting processes had failed.[16]

The NAO audit looked only at the performance of UKSV in processing clearances and renewals. It did not assess how effective those processes were at mitigating insider risk and protecting national security. An emphasis on performance metrics like speed and cost, as distinct from security effectiveness, is a common phenomenon. The NAO report noted that the frustrated customers of UKSV mostly expressed concerns about the *speed* of vetting decisions, not their quality. This focus on speed should be a cause for concern. Speed and quality in personnel security are not mutually exclusive, but there can be tension between them. Organisations that care only about the efficiency of their personnel security might inadvertently neglect its effectiveness, putting them at risk.

Case history

2022: UK airlines blamed delays in government national security vetting checks for holding up the recruitment of new airport staff, leading to chaos at airports as hundreds of flights were cancelled.[17]

Comment

• Delays in government vetting have real-world consequences outside government.

There are reasons to suspect that traditional (untransformed) government vetting might be less effective than it should be. It still relies on checking records periodically, often with large gaps in between. The current UK rules permit the lower levels of vetting clearance to remain valid for ten years before they must be fully renewed, and even the highest (DV) clearance is valid for seven years. Ten years is far too long. People and their circumstances change significantly over a decade. From the candidate's perspective, the vetting process can feel awkward and off-putting, making it a potential barrier to diversity and inclusion.[18] Furthermore, pressure to reduce backlogs has diverted resources away from the crucial in-trust (aftercare) measures that should be applied to existing employees and contractors. Considering that most insiders become insiders after they are recruited, aftercare is even more important than pre-employment screening.

In government departments whose entire workforce is vetted, a person whose clearance is withdrawn could lose their job. This can make colleagues less willing to report concerns, especially if they suspect the official response may be heavy-handed. The ubiquity of vetting in a high-security organisation can also foster a degree of complacency if people believe they are working in a secure environment where everyone is proven to be trustworthy.

The problems with national security vetting are not unique to the UK. A series of reports by the US Government Accountability Office (GAO) highlighted comparable concerns about US government vetting. In 2018, the GAO placed the government-wide personnel vetting process on its High-Risk List because of concerns about clearance backlogs, ageing IT systems, and uncertainties about the quality of investigations.[19]

Societies change over time and so too do the risk factors for insider behaviour. For national security vetting to remain relevant, its criteria must evolve. For example, homosexual acts were once regarded as a risk factor by US and UK government vetting authorities because of concerns about vulnerability to blackmail, but this has long since ceased to be the case. Attitudes towards personal finances have also shifted over time. A 2021 study of US government vetting found that the most common reasons for denying clearance were finances, personal conduct, and foreign influence. The study also identified four areas of emerging security concern:

- Growing opportunities for contacts with foreign nationals, facilitated by the Internet, globalisation, and study overseas
- Financial risks arising from student debt and the use of cryptocurrencies
- Increasing use of drugs such as analgesics and stimulants
- Risky online behaviour, such as sharing personal information on social media and accessing illegal content[20]

At the time of writing, personnel security in UK policing appears to be in need of major reform. A 2022 report by the official inspectorate of policing made dismal reading, with the senior inspector commenting that it was 'too easy for the wrong people both to join and to stay in the police'.[21] The report was commissioned following the 2021 murder of a young woman by a serving police officer (see Chapter 1). The inspectorate found that police vetting systems were inadequate, allowing unsuitable people to join police forces and continue serving there, despite evidence of their unsuitability. The consequences included misogynistic and predatory behaviour towards women, as well as corruption. Some forces had failed to make the connection between officers sexually harassing female colleagues and behaving similarly towards members of the public.

The report found that forces had for many years ignored warning signs that their vetting arrangements were not working effectively and had repeatedly failed to implement recommendations to improve them. Pre-trust measures were often inadequate, with some forces recruiting new officers without checking their references or interviewing them. Forces had employed officers with criminal records or known links to organised crime, substantial debts, or histories of misconduct and complaints in other forces. Applicants had got away with providing incomplete or false information. In some cases, recommendations to reject unsuitable candidates had been overruled for no apparent reason, except perhaps the pressure to boost officer numbers following years of cutbacks. External demands to meet recruitment targets had apparently caused some forces to increase their 'risk appetite'.[22]

Nationality

One of the insider risk indicators mentioned at the start of this chapter is association with a hostile foreign state. This raises the delicate issue of nationality. Public sector organisations, businesses, and academic institutions sometimes face an awkward dilemma when deciding whether to employ or work closely with someone who is a citizen of a hostile foreign state or has familial links to them. Organisations naturally want to benefit from an unfettered exchange of people and ideas. They also rightly want to avoid doing anything that might be discriminatory, unethical, or unlawful. However, they must balance these important considerations against the potential security risks.

There are valid reasons why a person's connections with a hostile foreign state might be a legitimate concern for a potential employer. The first, and least likely, is that the individual might already be acting covertly on behalf of the foreign state and would become an active insider if brought onboard. A more plausible possibility is that a hostile foreign state could put pressure on the individual or their family to cooperate once they are on board. Russian, Chinese, and Iranian state agencies are known to do this. The risk is not speculative. By allowing them in, the organisation might put itself at risk. Importantly, the organisation might also put the individual and their family at risk by making them targets for coercion. Responsible organisations have a duty of care to consider such possibilities.

Case history

2018: Estonian army major Deniss Metsavas was arrested by the Estonian authorities and charged with spying for Russia. He pleaded guilty and was given a lengthy jail sentence. Metsavas, an ethnic Russian, was recruited – or rather, coerced with blackmail – by means of a honeytrap while visiting relatives in Russia in 2007.[23]

Comments

- An individual's nationality and family connections with a hostile foreign state could be instrumental in their recruitment by an external threat actor.
- An insider may be recruited coercively, starting with a honeytrap. Yes, they do still happen.

Open-source intelligence (OSINT)

Businesses, universities, and other non-governmental organisations do not have ready access to the full range of official records available to governments. They must therefore turn to other sources of information. Many are making increasing use of open-source intelligence (OSINT) – the blanket term for gathering and analysing publicly available information. At its most basic, this amounts to searching the web, inspecting the individual's social media sites (with their permission), and consulting online databases.

More advanced OSINT methods use powerful software tools to 'scrape' all the information about the person of interest that is accessible via the Internet, including searches of sites that are not indexed by conventional search engines. OSINT tools have developed to a level where their capability is not far short of some classified methods used by government agencies. Commercial agencies are using powerful OSINT techniques to analyse security risks connected with hostile states like Russia and China – an area that was previously hard to investigate without access to secret intelligence. Indeed, newer OSINT techniques could help to strengthen traditional government vetting processes.

There is a fine line, however, between checking publicly available information and intrusive or even unlawful surveillance, especially if it is done without the individual's informed consent. Moreover, while it is relatively easy to amass large quantities of data about a person of interest, it is harder to make sense of the data and derive valid conclusions about that person's trustworthiness and likely future behaviour.

Psychometric testing

Psychometric tests are measurement tools designed by psychologists to assess specific features of an individual's psychological makeup or behaviour. Like any scientific measurement tool, they must be proven to be valid and reliable, which means they must demonstrably measure the things they are supposed to measure, and they must do so consistently and repeatably.[24] Psychometric tests vary in their validity and reliability. Some are better measuring instruments than others.

The right psychometric tests, applied in the right way at the right time, can enhance personnel security to a modest extent. They may be applied during the recruitment (pre-trust) phase as part of the selection process, or as part of the in-trust regime, or both. Their purpose would be to assess trustworthiness or potential for harmful insider behaviour. As we saw in Chapter 4, there is evidence to link some personality characteristics to insider risk. The best known of these are the Dark Triad traits of psychopathy, Machiavellianism, and narcissism. Psychometric tests designed specifically to measure these traits are sometimes used as part of a personnel security system. However, several caveats should be borne in mind.

There are thousands of different psychometric tests, only some of which are directly relevant to insider risk. Tests can be costly to use if they are commercially copyrighted, and they may require specialist expertise to interpret the results. Some organisations routinely employ psychometric tests as part of their recruitment process, to help them judge suitability and organisational fit. However, many of these tests measure characteristics that have little direct bearing on insider risk or trustworthiness. Moreover, tests are to varying extents vulnerable to manipulation by applicants, some of whom misrepresent themselves to achieve a favourable outcome.[25] Research on screening for US government positions found that between a quarter and half of all applicants deliberately distorted their responses to conceal problems and make themselves look good.[26]

Table 7.1 Elements of pre-trust security

- Verify identity. Is this person who they say they are?
- Verify right to work. Is this person legally entitled to work in this country?
- Check credentials. Has this person told the truth about their education, employment history, residence, and travel? Are there unexplained gaps in the timeline?
- Check criminal records. Does this person have an unspent criminal record? If so, does it signify an unacceptable level of security risk or reputational risk? Have they told the truth about their criminal record?
- Check national security records (if permitted). Is this person known to have links to terrorism, hostile foreign state activity, or other threats to national security?
- Check for drug and alcohol abuse. Does this person use illegal drugs? Do they abuse alcohol or other drugs?
- Assess suitability and organisational fit. Does this person have the right skills and personal attributes to do the job? Will they be comfortable in this working environment and organisational culture? Consider using psychometric tests to assess relevant personality attributes and attitudes.
- Assess trustworthiness. Has this person been truthful about everything they have told us? Have they demonstrated integrity or lack of integrity in the past? Can we trust them? Consider applying the following methods:
  - Structured security interview
  - OSINT searches and analysis
  - Psychometric tests of personality attributes associated with insider risk (e.g. Dark Triad)

A further potential problem with psychometric tests, and psychological indicators more generally, is that the correlations with insider actions are generally quite weak. In scientific terms, the effect size is small. The psychological characteristics that are more closely associated with insider risk, such as narcissism, are also common in the general population, and most people who display these characteristics do not become active insiders. Consequently, the tests are likely to generate many false positives.

The key elements of pre-trust security are summarised in Table 7.1.

Discussion points

- How would you conduct an interview to assess the trustworthiness of a recruitment candidate?
- How would you decide whether to employ someone with a criminal record?
- How would you decide whether to employ someone who is a national of a hostile foreign state?
- What can a person's finances tell you about their trustworthiness?
- Describe a good interview and a bad interview.
- Should recruitment and personnel security be combined in a joint function?

Notes

1 CIPD. (2015). *A Head for Hiring: The Behavioural Science of Recruitment and Selection.* London: CIPD.
2 Ackerman, J. M., Nocera, C. C., and Bargh, J. A. (2010). Incidental haptic sensations influence social judgments and decisions. *Science, 328*: 1712–1715.
3 Vrij, A., Hartwig, M., and Granhag, P. A. (2019). Reading lies: Nonverbal communication and deception. *Annual Review of Psychology, 70*: 295–317.
4 Vrij, A. et al. (2022). Verbal lie detection: Its past, present and future. *Brain Sciences, 12*: 1644.
5 Vrij, A. and Fisher, R. P. (2016). Which lie detection tools are ready for use in the criminal justice system? *Journal of Applied Research in Memory and Cognition, 5*: 302–307.
6 Palena, N. et al. (2021). The Verifiability Approach: A meta-analysis. *Journal of Applied Research in Memory and Cognition, 10*: 155–166.
7 Vrij, A. et al. (2022). Verbal lie detection: Its past, present and future. *Brain Sciences, 12*: 1644.
8 ACFE. (2022). *Occupational Fraud 2022: A Report to the Nations.* https://legacy.acfe.com/report-to-the-nations/2022/
9 HMICFRS. (2022). *An Inspection of Vetting, Misconduct, and Misogyny in the Police Service.* November 2022. www.justiceinspectorates.gov.uk/hmicfrs/publications/an-inspection-of-vetting-misconduct-and-misogyny-in-the-police-service/
10 BBC. (2023). www.bbc.co.uk/news/uk-england-lancashire-64797676
11 CAA. (2022). www.caa.co.uk/news/commercial-pilot-sentenced-for-fraud/
12 BBC. (2021). www.bbc.co.uk/news/uk-england-shropshire-57603091
13 BBC. (2021). www.bbc.co.uk/news/uk-england-london-56941544
14 MoJ. (2010). Conviction histories of offenders between the ages of 10 and 52. *Ministry of Justice Statistics Bulletin*, 15 July 2010. https://assets.publishing.service.gov.uk/government/uploads/system/uploads/attachment_data/file/217474/criminal-histories-bulletin.pdf
15 Scott, P. F. (2020). The contemporary security vetting landscape. *Intelligence and National Security, 35*: 54–71.
16 NAO. (2023). *National Audit Office: Investigation into the Performance of UK Security Vetting.* HC 1023. London: Cabinet Office.
17 Middleton, J. (2022). Airline trade body blames recruitment approval delays for UK airports gridlock. *The Guardian*, 31 May 2022.
18 Lomas, D. W. B. (2021). "Crocodiles in the corridors": Security vetting, race and Whitehall, 1945–1968. *Journal of Imperial and Commonwealth History, 49*: 148–177.

19 GAO. (2021). *Personnel Vetting. Actions Needed to Implement Reforms, Address Challenges, and Improve Planning*. GAO-22-104093. December 2021. www.gao.gov/assets/gao-22-104093.pdf

20 Posard, M. N. et al. (2021). Updating personnel vetting and security clearance guidelines for future generations. RAND Research Report RRA757-1. Santa Monica CA: RAND Corporation. www.rand.org/pubs/research_reports/RRA757-1.html

21 HMICFRS. (2022). *An Inspection of Vetting, Misconduct, and Misogyny in the Police Service*. November 2022. www.justiceinspectorates.gov.uk/hmicfrs/publications/an-inspection-of-vetting-misconduct-and-misogyny-in-the-police-service/

22 Ibid.

23 Weiss, M. (2019). The hero who betrayed his country. *The Atlantic*, 26 Jun 2019.

24 Bateson, M. and Martin, P. (2021). *Measuring Behaviour: An Introductory Guide*. 4th edn. Cambridge: Cambridge University Press.

25 Baweja, J. A., Burchett, D., and Jaros, S. L. (2019). An evaluation of the utility of expanding psychological screening to prevent insider attacks. OPA Report 2019-067. PERSEREC, US Department of Defense. https://apps.dtic.mil/sti/pdfs/AD1083812.pdf

26 Levashina, J. (2018). Evaluating deceptive impression management in personnel selection and job performance. In *Clinical Assessment of Malingering and Deception*, 4th edn., ed. by R. Rogers and S. D. Bender. NY: Guilford.

8   In-trust measures

Reader's Guide: This chapter describes the in-trust personnel security measures (also known as aftercare) that can be applied after a person has been trusted with access to an organisation's assets.

In-trust personnel security measures are intended to protect an organisation from individuals who have been brought on board and trusted with access. In-trust measures are sometimes referred to collectively as aftercare or ongoing security.

As noted before, the evidence from known insider cases shows that most active insiders develop their insider intentions after joining their organisation. The archetypal insider is an employee or contractor who has worked for the organisation for a few years. The clear implication is that no organisation can afford to rely too much on pre-trust measures. Most of their insider risk will stem from trusted individuals who have passed through pre-employment screening. And yet, many organisations do depend heavily on pre-trust measures and pay insufficient attention to what happens afterwards, leaving them exposed to unrecognised and unmitigated insider risk.[1]

The main categories of in-trust personnel security are:

- Access controls
- Exit controls (e.g. data loss prevention)
- Behavioural controls
- Awareness-raising, training, and communication
- Reporting channels and management oversight
- Automated monitoring
- Investigation
- Sanctions (e.g. disciplinary measures)
- Exit procedures

Access controls

The purpose of access controls is to mitigate insider risk by limiting people's legitimate access to the organisation's assets. An active insider can inflict maximum damage if they have unrestricted access to data, money, intellectual property, infrastructure, buildings, and people. Physical access controls are used for limiting access to places, while virtual access controls limit access to data and digital systems.

A basic tenet of protective security is the principle of least access, which states that a person (or other entity) should only have access to the data and other assets they need to do their job. In the intelligence world, the principle is sometimes referred to as need-to-know. The principle is

DOI: 10.4324/9781003329022-11

sound in theory, but it runs counter to a predominant culture of sharing and collaboration, which tends to maximise access. Applying the principle of least access entails some trade-off between security and ease of conducting normal business.

Case history

1983: MI5 officer Michael Bettaney covertly volunteered his services to the Soviet KGB intelligence service in London during the height of the Cold War (as mentioned in Chapter 3). At the time, Bettaney was a middle-ranking officer in the MI5 counter-espionage section responsible for investigating Soviet intelligence activity in the UK. His betrayal was secretly reported to British intelligence by Oleg Gordievsky, a KGB officer stationed in London. Bettaney's job was to investigate Gordievsky and his KGB colleagues, but he did not know that Gordievsky was a British agent. Following the tip-off, Bettaney was arrested and jailed.[2]

Comment

• Gordievsky's secret role as a British agent was protected by rigorous application of the need-to-know principle within MI5. Had Bettaney known of Gordievsky's agent status, he would almost certainly have betrayed him to the Soviets.

Virtual access to digital systems and data is controlled through technical measures such as multi-factor authentication, which requires users to prove their credentials in at least two different ways before they can gain access. Other technologies employ micro-behavioural cues, like the individual's distinctive way of using a mouse or a keyboard, to discriminate between legitimate users and threat actors who have stolen their log-on credentials.

So-called zero-trust cyber security systems take the principle further by repeatedly requiring users and devices to re-authenticate their credentials. They function as though the outer digital perimeter has been breached and nothing inside the system can be fully trusted. The aim is to make it harder for an external attacker or active insider to move around freely within the system. Zero-trust cyber security is analogous to a physical security regime in which people must display their security pass and key-in the associated PIN when moving within the physical perimeter of a secure building or facility.

In high-security organisations, additional layers of physical and cyber security controls are used to restrict access to the most sensitive information. Rigorous technical and physical controls can limit access to cyber systems and buildings with a high degree of assurance, while encryption can provide additional protection for data at rest or in motion. But none of these measures can fully protect against a trusted insider abusing their legitimate access.

Compartmentalisation of assets

The ability of access controls to reduce security risks can be enhanced by dividing assets into separate subsets, each of which has its own access controls. This concept is more commonly applied to the protection of data, where it involves dividing databases into multiple blocks, or compartments. If an external threat actor or insider enters one part of the network, they do not automatically get access to the entire network and all the data. Provided the controls are effective, an insider's ability to cause harm is contained within those compartments to which they have access. Compartmentalisation can slow down an attack and improve the likelihood of detection. Determined insiders may of course find illicit ways of extending their access. Edward Snowden, who was mentioned in Chapter 1, is an example.

In the physical world, the equivalent of compartmentalisation is dividing a building or estate into secure enclaves, each with its own physical access controls. This should reduce the chances of an intruder enjoying a free rein if they manage to penetrate the outer perimeter. In some government buildings, there may be a SCIF (Sensitive Compartmented Information Facility) where only authorised individuals with the requisite clearances can access highly classified material. Again, these secure enclaves will not exclude insiders who have legitimate access.

Structural safeguards

Another way of augmenting protection against insiders is by imposing procedural or engineering barriers that make it necessary for more than one person to conduct an illicit insider action. The best-known example is the dual-key safeguard against the unauthorised launch of a nuclear weapon, whereby at least two key-holders must independently authorise a launch. This arrangement, which is built into the hardware, makes it physically impossible for a lone insider to fire a weapon. Comparable methods are used in the civil nuclear industry to prevent lone individuals from gaining unauthorised access to radioactive materials. Engineering safeguards make it impossible for one person acting alone to remove dangerous radioactive material from its secure storage. Similarly, banks may require two members of staff to cooperate before a high-security vault can be opened.

Of course, it is possible that two or more insiders could conspire to cause harm. A significant proportion of insider fraud cases do involve two or more perpetrators.[3] Or, as in the Northern Bank robbery described in Chapter 4, two reluctant insiders could be coerced into cooperating under threat of violence. Other things being equal, however, the more people it takes to form a conspiracy, the less likely it is to succeed. Conspiracies do happen, but most insider attacks are conducted by lone individuals.

Role-based security

We saw in Chapter 3 that one of the variable characteristics of insiders is the extent to which their role gives them legitimate access. On average, roles with more access, like senior managers, IT systems administrators, or security personnel, pose a bigger potential risk than those with less access.

More sophisticated personnel security systems take account of this relationship between access and risk with a methodology called role-based security. This involves identifying the roles with the greatest access and applying more stringent personnel security measures to them. Paying more attention to the biggest potential risks is sensible. However, role-based security does have limitations:

- People often have more access than is apparent from their job title.
- Few organisations operate strict need-to-know barriers. In practice, the demarcation between high-risk and low-risk roles is rarely watertight.
- People tend to acquire more access over time as they move around an organisation. In cyber security, the accumulation of access rights by IT users is known as privilege creep.
- Determined insiders will find illegitimate ways of extending their access beyond the formal boundaries of their role.

- Access is a continuous variable that is hard to measure objectively for many roles. Dividing all roles into a few broad categories (e.g. 'high risk' and 'low risk') is a huge simplification, though still better than treating them all the same.
- Identifying high-risk roles and controlling them more strictly can induce a false sense of security. At worst, it fosters complacency about all the other roles. Even a supposedly low-risk role can present a substantial risk if the person occupying it is a determined insider. The *role* might be low risk, but the person might not. Furthermore, a large number of low risks can add up to a large risk.
- Maintaining role-based security requires administrative processes to keep permissions up to date and manage exceptions, such as when a user has a legitimate need to access assets outside their defined role. This takes time and money.

Various technology tools are available for automatically enforcing and monitoring the access privileges of IT system users. They are sometimes referred to as Privileged Access Management (PAM) tools. Incidentally, the use of physical and cyber security access controls to enhance personnel security is an example of the holistic approach described in Chapter 6.

Cyber security professionals have coined the term Very Attacked Persons (VAPs) to denote people within their organisation who attract a high frequency of targeted cyber and phishing attacks. VAPs typically include senior leaders, IT systems and financial administrators, and cyber security professionals. What they have in common is juicy access.

Exit controls

Exit controls are intended to stop insiders or other threat actors from removing valuable assets to which they have had access. The two main types of exit control are:

- Data loss prevention (DLP) technologies to prevent the unauthorised removal or exfiltration of data.
- Exit searching at buildings to deter and prevent people from removing valuable physical assets such as sensitive documents, IT hardware, or saleable items.

DLP technologies are common in larger organisations, especially those subject to regulation. In its simplest form, DLP works by automatically applying a set of rules that block specific actions, such as emailing large files to an unknown external address.

Exit controls and access controls provide metaphorical bookends to the protection of assets. Access controls prevent insiders from having unnecessary access to assets, while exit controls prevent them from illicitly removing those assets.

Behavioural controls

The most reliable access and exit controls are embodied in the technology or physical infrastructure, making it difficult or impossible for people to perform certain actions that are forbidden. However, such controls can only protect against specific types of prohibited actions. They also cost money and add friction to everyday business. It is therefore tempting to rely mainly on so-called behavioural controls, which in essence means telling people not to do bad things. So, for example, rather than using a technical control to block IT users from sending files to personal email addresses, a behavioural control would rely on instructing people not to do it.

The obvious weakness of behavioural controls is that people do not always do what they are told. They might break a rule deliberately because they are an intentional insider. Or they might do so for other reasons – for instance, because they are unaware of the rule, do not understand the rule, forgot the rule, do not care about the rule, or made an honest mistake because they were distracted or tired. If the consequences of violating a security policy would be very serious, then an organisation should try to find technical or physical controls that make the action difficult or impossible, rather than simply issuing instructions and hoping everyone will comply.

The treatment of phishing emails is a case in point. A behavioural control would involve instructing IT users not to click on suspicious attachments that might contain malware. The instruction might be accompanied by online training to improve people's ability to distinguish between phishing emails and genuine emails, possibly reinforced with a threat of sanctions for those who err. In real life, however, even the most conscientious people make occasional misjudgements. Well-crafted phishing emails are hard to spot, and the best ones can fool cyber security professionals. Furthermore, some users will knowingly click on suspicious-looking attachments because they are reckless and do not care about the consequences. With the best will in the world, an organisation should expect at least five per cent of phishing emails to be clicked through, no matter how many awareness campaigns it runs. A large-scale study of simulated phishing attacks in 2021 found that users clicked on 20 per cent of attachments and 11 per cent of links.[4] And remember, it only takes one click to infect a network. A better approach is to combine behavioural controls with technical controls that help users to identify suspicious emails and stop them from behaving recklessly.

The organisational mindset that relies heavily on behavioural controls is epitomised by the security cliché that people are the weakest link. Pinning the blame on people is often an excuse for not tackling the root causes, which are bad policies, faulty processes, and inadequate technology.

Awareness-raising, training, and communication

Everyone has a role to play in defending their organisation against insider risk. Good personnel security relies on people understanding the risk and knowing what to do about it. The standard approach to achieving this is through awareness-raising and training, which should be tailored for different audiences. The aims should include:

- Equipping the workforce and leaders with a sufficient understanding of insider risk, its symptoms, causes, and consequences.
- Explaining why any responsible organisation should apply proportionate security measures to protect itself and its people.
- Explaining why everyone in the organisation has some responsibility for security.
- Stating in clear practical terms what people should do if they notice something that does not look right.

These aims can be pursued in various ways, including through internal communications, online training sessions, table-top exercises, red-teaming exercises, simulations, deep-dives into insider cases, regular reporting of relevant metrics, and tailored briefings to leaders, managers, and specialists in related functions like HR and cyber security.

People are more likely to notice potentially significant changes in a colleague's behaviour if they know that person. The frequent movement of staff can therefore be a barrier to the detection and reporting of warning signs. Internal reorganisations, staff shortages, and dysfunctional HR processes can generate a rapid turnover of people, leaving teams composed of relative strangers. Excessive churn weakens one of the best mechanisms for detecting insider risk. This problem has been compounded by the shift to remote working that was accelerated by the Covid-19 pandemic. It is harder to keep a friendly eye on a colleague if you seldom or never meet them in person.

Top-down communication from leaders can help if it raises awareness and reinforces a positive security culture. However, the wrong kind of top-down communication can make matters worse if it creates an authoritarian climate in which people are frightened of speaking up. Lateral communication between workers may reinforce or undermine the messages from leaders, depending on how they regard those leaders.[5] Leadership and culture are discussed in Chapter 9.

Reporting channels and management oversight

A key objective of personnel security is to detect, interpret, and triage the weak early signals of potential insider risk so that targeted action can be taken to prevent the risk from developing. The best detectors of these signals are people – the potential future insiders themselves, their colleagues, and their managers.

A crucial element of any personnel security system is the set of reporting channels through which individuals can communicate any concerns they might have, whether about themselves or their colleagues. There is good evidence that these channels make a difference. For example, a study of insider fraud found that the commonest way in which frauds were detected was by tip-offs from employees. Organisations that had hotlines or other reporting channels detected frauds sooner and suffered smaller losses. The losses were on average twice as high in organisations without reporting hotlines.[6]

Organisations should not take reporting for granted. Creating hotlines and other reporting channels does not guarantee that people will use them. Employees do not always speak up. The reasons why people do, or do not, report concerns are complicated. Unsurprisingly, two of the main inhibitors are believing that speaking up would be futile, and a fear of adverse consequences, such as causing problems for colleagues or being viewed as a troublemaker. Disgruntlement is another inhibitor; people who feel ill-disposed toward their organisation are less inclined to help by pointing out its problems. People are also unlikely to report if they do not know how to do so, or do not trust the organisation to use the information discreetly. A study of one large organisation found that people were less likely to report concerns about a potential insider if they had doubts about the confidentiality of the reporting process, or if the process was unclear.[7] Organisational leaders are often unaware of the extent to which people remain silent. They may delude themselves that an absence of bad news is good news.

The ideal users of reporting channels would be the individuals who are experiencing difficulties, and who therefore might be on the developmental path to insider action. As noted in Chapter 6, the best way of forestalling insider risk is with a welfare approach, by which individuals are helped to resolve problems that might otherwise fester. Most people with personal or work problems are not on the path to insider action, and punishment is rarely the right first move. Accordingly, the best personnel security systems have multiple discreet channels through which individuals can surface their problems in the confident knowledge that they will receive a reasonably sympathetic response and will not be penalised for coming forward.

If the possible future insiders themselves do not report their incipient problems, then the next line of defence is their colleagues and managers. They should be encouraged to feel responsible for supporting security and empowered to report concerns about individuals who display signs of possible insider intent. But what *are* those signs, and what should colleagues and managers be looking for? For the system to work, bystanders must notice the relevant behaviour and recognise its significance. The most salient risk indicators are described later in this chapter. However, any reporting system must allow people to exercise their judgement and act if something does not look right. It should not be reduced to a prescriptive list of specific warning signs.

Before individuals go on to commit damaging insider actions, they may display so-called counterproductive work behaviours, such as arguing with colleagues, performing their job poorly, frequently being late for work, or infringing minor rules. Researchers in the UK found that organisational change, resulting from such things as restructuring or mergers, is one of the main drivers of counterproductive work behaviour. If left unchecked, it can develop into harmful insider action. This sort of behaviour is usually noticed by colleagues and managers, but it is rarely reported as an indicator of something potentially more serious.[8] This lack of reporting may not be entirely a bad thing, as many people display some of these behaviours at times, but only a small minority become active insiders. If every morose, hungover, cantankerous, or underperforming person were to be reported as a potential insider, organisations would be overwhelmed by false positives.

Automated detection and monitoring

Technology is making large and rapidly growing contributions to personnel security. As mentioned earlier, DLP technologies can stop insiders from exfiltrating data through unauthorised channels. Other automated systems can help in detecting insider risk and managing its consequences.

User Activity Monitoring (UAM) tools monitor and record the online activities of IT system users. They can support personnel security by detecting potential indicators of insider activity and by informing investigations. Security Information and Event Management (SIEM) tools collect, aggregate, and analyse log data from devices and networks. They can help to detect insider activity by spotting anomalies and supporting investigations. The functionality of SIEM tools may be extended with machine learning (ML)-enabled analytics tools to identify patterns in complex datasets. They are sometimes referred to as User Behaviour Analytics (UBA) tools. Finally, if the risk materialises and an insider breaches cyber security, digital forensics tools can assist investigators in pinpointing what happened and who did it. Products are available that combine all these tools in a single package.

Numerous commercial software tools purport to detect the signs of insider activity. They work by searching through large volumes of data for actual or incipient insider actions, anomalous behaviours, or violations of security policy. It is worth distinguishing here between three functions that these systems might perform: detecting unusual behaviour that might be a sign of insider intent (so-called anomaly detectors); detecting known precursors of insider action (attack pattern tripwires); and detecting actual violations of policy or other clearly transgressive actions (policy violation tripwires).[9]

The types of IT user behaviours that are easiest to detect reliably are the clear policy violations, i.e. unequivocal actions that are explicitly forbidden by security policies. Examples might include accessing prohibited websites, downloading unauthorised software from the Internet, unauthorised copying of files to removable storage media, copying or removing large or sensitive datasets, interfering with system software or audit logs, excessive printing, logging in from two different geographical locations at the same time (implying misuse or compromise of credentials), or sending corporate data to personal email accounts. Detecting these actions should help to protect IT systems and data, but it clearly would not deal directly with other forms of insider action. As we have seen, insiders do more than violate cyber security protocols.

Moreover, finding such violations is more about dealing with the problem than preventing it from developing. An insider who purposefully breaches IT policies might already be causing more serious harm elsewhere.

More sophisticated automated monitoring systems use artificial intelligence (AI) or ML to identify anomalous events or behavioural risk indicators on IT systems.[10] They offer the alluring prospect of detecting weak precursor signals and hence the opportunity to intervene at an earlier stage. But such systems are expensive and require expertise to operate effectively. Their real-world performance does not always match their vendors' claims. As we shall see, they can also raise ethical issues and produce unintended consequences.

It is difficult for any automated system to discriminate accurately and reliably between normality and anomalies, and even harder to pinpoint the very few anomalies that accurately predict insider actions. Behavioural anomalies must be defined in relation to some baseline definition of normal. The 'normal' comparator could be the individual's previously established pattern of online behaviour, or the typical behaviour of everyone in a similar job role, or the typical behaviour of everyone in the workforce. They will generate different answers. Anomaly detectors tend to have high false-positive rates, though they are getting better.

Building a DLP tool that blocks specific actions is relatively straightforward, but detecting the subtle early warning signs of a potential future insider is far harder. Capable insiders conceal their intentions and use their inside knowledge and access to avoid arousing suspicion. Their covert actions are inherently difficult to detect. Automated systems might identify interesting signals, like unusual working patterns or sudden changes in online behaviour, but the correlations between such signals and insider intention are often weak or unproven.

Conventional ML methods struggle to differentiate between potential insiders and everyone else. There are several reasons for this, including the complexity of the data, the lack of validated risk indicators, and the adaptive nature of the risk. Adaptive risks are particularly challenging for conventional ML systems that must be re-trained on new methods of attack. An ideal system would respond adaptively to changes in the risk without the need for re-training.

Deep-learning methods offer better prospects for finding meaningful patterns within the morass of data. These systems learn from complex raw data with minimum human intervention. They have proved hugely successful in applications like speech recognition, computer vision, and natural language processing.[11] They are good at sifting through vast quantities of complex data to find patterns and anomalies. However, no current or easily foreseeable technology has the mind-reading capability needed to decipher the private intentions of a covert insider.

A practical problem for ML-based methods is the scarcity of real-world data with which to train the systems or test their accuracy. There is a severe shortage of reliable data about confirmed insider incidents and the sequences of events leading up to them. Synthetic data sets are often used as substitutes for real-world data. Their quality and realism are variable. One of the best sources is the database of real-world insider cases maintained by Carnegie Mellon University.[12] This has been used to generate synthetic datasets of insider cases for use in research. However, when the same datasets are used repeatedly in different research studies, as is often the case, the strength of their collective findings is reduced. Multiple simulations or experiments that use the same dataset are not statistically independent of one another, which makes the totality of their conclusions less than the sum of its parts.

In addition to being in short supply, real-world training data does not include data about those skilful or lucky insiders who have not been discovered. Automated systems that have been trained on a sample of known cases might therefore be biased against discovering the most successful and damaging insiders. To be fair, though, the same could be said for human investigators, who similarly do not know what they do not know.

In another unwelcome parallel with humans, the current generation of generative AIs are prone to fabricating stories, or 'hallucinating'. In amongst their remarkably accurate outputs, they sometimes produce detailed answers that look plausible but turn out to be inaccurate or completely untrue.

People are currently still the best detectors of insider risk, provided they know what is expected of them. Evidence from US government research has shown that behavioural indicators, of the sort that can be observed by people, are detectable at an earlier stage than technical indicators such as unusual patterns of online behaviour. Therefore, organisations that rely exclusively on technical indicators may not become aware of an insider risk until it is too late. Using only cyber-technical tools to protect against insiders has been likened to a drunk person searching for their lost keys under a streetlamp; they search there because that is where the light is, not because that is where they dropped their keys.[13] If the goal is prevention rather than cure, a better strategy is to use a combination of behavioural and technical indicators, with machines and people working together.[14]

The dangers of intrusive monitoring

Automated monitoring is an established practice in some industries. Financial regulators require banks to monitor their traders for compliance with insider trading rules. Traders must accept that their phone calls, emails, and online transactions are routinely recorded and analysed. People working in call centres are monitored for their performance in dealing with calls, and there has been a growth in the use of software tools for checking the productivity of people working from home. Similarly, employees of intelligence services, the military, and other high-security organisations must accept that a degree of monitoring comes with the job, along with more intrusive personnel security and tighter restrictions on their behaviour. People who choose to work for these organisations generally acknowledge that such impositions are necessary and proportionate in view of the risks, including the risks to their own safety and livelihoods. In most other contexts, however, automated monitoring and intrusive surveillance raise significant issues of trust, privacy, and legality. They can also do more harm than good.

In some countries, employers are permitted by law to monitor their employees' use of corporate IT systems, although they are not permitted to monitor behaviour more generally. In addition to being lawful, such monitoring may be regarded as acceptable within the prevailing culture of an organisation. But even if monitoring is both lawful and culturally acceptable, it still might not be the right thing to do. The wrong kind of monitoring can have unintended consequences, including making insider risk worse.

People who work remotely report that being monitored makes them feel anxious and resentful about the implied lack of trust. Research has consistently found that such surveillance can have a range of adverse effects. These include undermining morale, breeding resentment, lowering productivity, and eroding loyalty to the organisation. Intrusive monitoring can also suppress creativity by deterring individuals from behaving in novel ways that might attract adverse attention. It can make employees less respectful of security measures and more likely to break rules. At worst, it may have the perverse effect of provoking individuals to commit insider actions.[15]

A European Commission study which reviewed almost 400 published articles found that excessive monitoring produced a variety of negative effects, including increases in stress and counterproductive work behaviour, reductions in job satisfaction and commitment to the organisation, and negative feelings about privacy, trust, and autonomy. The study highlighted the dangers of mission creep. By monitoring access control systems, webcams, emails, and keystrokes, employers could gather information about their employees' thoughts, emotions, locations, and movements, as well as their task performance. The report noted that the surveillance of workers in Europe, the UK, and the US had intensified during the Covid-19 pandemic, eroding the boundaries between work and home life.[16] Monitoring can feel very intrusive, especially if its purpose is unclear. It also sends people a message that their employer does not trust them. As we have seen, distrust breeds distrust.

The false positives generated by automated monitoring systems are pernicious. People rightly feel aggrieved if they or their colleagues are subjected to unjust suspicions of wrongdoing based on faulty analysis of flimsy evidence.[17] If an advertising agency uses automated sentiment analysis to predict who might be interested in buying their product, a high incidence of false positives does not matter much. So long as the tool gives them a commercial edge, it pays to use it. Very different considerations apply when using automated systems to predict who might be an insider, where even small numbers of false positives could cause serious upset. The tolerance for error should be much lower in the personnel security context than in marketing applications. The adverse effects of monitoring are worse if the organisation is not open and honest about what it is doing.[18]

A further concern with automated monitoring systems is their vulnerability to being gamed. Sooner or later, smart people will deduce or discover the behaviours that the system is looking for, enabling them to subvert the system by deliberately deceiving it and creating a more favourable impression. Suppose, for example, that a system was searching for changes in the linguistic style of users' emails, like shifting from the collegiate 'we' to the supposedly more alienated first-person 'I', or from mainly positive sentiments to mainly negative sentiments.[19] Even supposing such behaviours were valid indicators of insider risk, once the search criteria became known a user could adapt their communication style to avoid being pinged as a potential insider. People who are monitored while working remotely have increasingly found ways of deceiving the software by making themselves appear more productive than they really are. A determined insider could circumvent monitoring altogether by remembering crucial information and writing it down later.

Case history

1986: Ronald Pelton, a former intelligence analyst in the US National Security Agency (NSA), was convicted of spying for the Soviet Union and jailed for life. Pelton resigned from the NSA in 1979 and volunteered his services to the Soviets in 1980. Over the following years he gave the Soviets large amounts of highly classified information and advice about US intelligence operations, for which he was paid. Pelton had no access to classified documents during this time and relied on his memory. He was eventually uncovered following the defection to the US of a Soviet intelligence officer.[20]

Comments

- No automated monitoring system could prevent a knowledgeable insider from revealing secrets stored in their memory.
- Another example of an insider whose harmful actions started after they left their organisation.
- Another example of a spy who was only discovered following a defection from the other side.

Despite its pitfalls, the right technology can assist human decision makers by furnishing them with timely prompts and actionable information. Automated decision-support systems, as their name implies, support humans but do not replace them. It has been suggested that we should think of automated systems not as mere tools to help human decision makers, but rather as teammates in a collaborative venture to solve difficult problems. The different strengths of humans and machines combine to produce a system that is more than the sum of its parts.

The rapidly evolving technologies for automated monitoring, analysis, and decision support are bringing us closer to a prized goal of personnel security, which is to make it a continuous process. The aim of continuous evaluation (CE), as US authorities call it, is to review the trustworthiness of individuals in near real-time, rather than once every several years.[21] CE has been a gleam in the eye of western governments for decades and, at the time of writing, it remains work in progress. A CE system could triage data so that only those higher-risk individuals who were flagged by the system would receive closer scrutiny. The proponents of CE argue that it need not be more intrusive than current methods where everyone is reviewed periodically. Building CE systems that work is clearly not easy. One challenge is acquiring the streams of relevant behavioural and technical data. Another is proving that the system can extract valid risk indicators from these data streams.

Finally, it is worth noting that the distinction between automated decision-support tools and automated decision making is not clear-cut. Automated systems that triage and analyse information before serving it up to human decision makers are still making decisions, including decisions to disregard certain information. Moreover, busy people may come to rely so much on the tools that the machine's advice in effect becomes the decision. There might appear to be a human decision maker in the loop, but in practice it would be the machine making the decisions.

What should we look for? Indicators of insider risk

To stop insiders from causing harm, we need a combination of people and machines to spot the indicators of emerging insider risk. The question is: what *are* those indicators? The honest answer is that no one is quite sure. Good progress has been made in identifying behavioural and situational factors that correlate with insider risk, but the picture is far from complete.

Dozens of explanatory models, frameworks, and schemes for risk indicators have been published, but there is as yet no consensus on how best to identify an emerging insider risk. Some of these schemes only consider technical indicators of insider actions on IT networks, as though insider risk were merely a subset of cyber security. These cyber-centric models can be safely ignored for our purposes. Some other schemes present lists of behaviours that are thought to be signatures of insider risk but without empirical evidence to back them.

It might be tempting to believe that common sense is a sure guide to spotting insiders. However, common sense is not the same as valid evidence, and it is sometimes wrong. (According to Albert Einstein, common sense is the collection of prejudices we acquire by age eighteen.) A better approach is to analyse empirical evidence from known insider cases and identify behavioural indicators that are demonstrably indicative of risk (albeit with the proviso that they inevitably exclude indicators from unknown insiders).

A good example of an evidence-based list of behavioural indicators of insider risk is this one published by the US National Insider Threat Center:[22]

- Significant debt
- Living above one's means
- Bankruptcy

- Poor performance, complaints, absences
- Resignation
- Post-departure reach-back to current employees
- Unusual working hours
- Conflicts with colleagues or supervisor
- Previous rule breaking
- Violations of rules or policies
- Failure to complete security awareness training

In principle, these indicators of insider risk could be identified by colleagues and managers or some combination of people and automated systems.

It would be easy to add other plausible indicators to this list, even if their association with insider actions is not always proven. Examples that are often cited include: arrests or criminal convictions; drug or alcohol abuse; traumatic experiences; unusual travel; unsafe social media behaviour; overt dissatisfaction with work; mental health problems; stressful life events; poor performance appraisals (which are not the same as poor performance); abuse of expenses; and losing equipment.

A multitude of insider risk indicator schemes, ontologies, taxonomies, and frameworks are available to the confused personnel security professional.[23] Of these, the two most influential and widely used ones are the CERT framework produced by Carnegie Mellon University[24] and the SOFIT (Sociotechnical and Organizational Factors for Insider Threat) framework produced by Frank Greitzer and colleagues, also in the US.[25] Both are based on empirical data from known insider cases, mostly within US government and national security-related organisations.

SOFIT takes an evidence-based approach to identifying specific risk indicators that could be observed by people or by automated monitoring systems. SOFIT includes no fewer than 271 individual factors and 49 organisational factors. It provides an ontology, or taxonomic structure, for organising these many factors into categories. The two overarching categories are individual factors (e.g. job performance, psychological characteristics) and organisational factors (e.g. management, security arrangements). Unlike many other schemes, SOFIT is supported by empirical evidence. Even so, its validity has not yet been fully established. Further work is needed to demonstrate empirically that the indicators do reliably predict insider actions to a useful extent in a range of circumstances.[26] It would be good to know which indicators, or combinations of indicators, provide the best guides.

Many other risk indicator schemes refer to CERT and SOFIT and draw on the same US government data, so it is unsurprising that they have features in common. Most of the credible schemes involve some combination of individual characteristics (e.g. narcissism, erratic behaviour, mental health problems); workplace factors (e.g. dissatisfaction with work, toxic work environment, security violations); and other situational factors (e.g. financial problems, ideology, unusual foreign travel). You will notice that this list contains a mixture of observable behaviours (e.g. violating security rules), internal factors (e.g. mental health problems), and external factors (e.g. toxic work environment). Some of these schemes are fabulously complicated and liable to leave the bemused practitioner wondering what to do.

A couple of caveats are called for here. The first is that much of the evidence on which these schemes are based comes from US government and national security organisations. It cannot be assumed that these schemes are universally valid and work equally well for commercial organisations, small businesses, universities, or institutions outside the US. A second caveat is that they mainly deal with intentional (or 'malicious') insiders. They may not work as well with unintentional or unwitting insiders, for whom the risk indicators probably look quite different. For example, fatigue is a likely risk indicator for unintentional insider actions, because it makes people error-prone and impatient, but it seems less relevant to the long-term intentional insider. Similarly, a lack of awareness or understanding of security rules is a plausible risk factor for unintentional insider behaviour, but less so for the intentional.

Two other rules of thumb may be extracted from the blizzard of work in this area:

- *Look for combinations of factors*. No single risk factor works optimally by itself. Insider behaviour emerges from interactions between multiple internal and external factors. The best indicators of insider risk are therefore combinations of factors. This principle is embodied in the developmental model outlined in Chapter 4.
- *Look for new or unusual behaviours*. There are grounds for thinking that personnel security systems should pay particular attention to behaviour that is new or unusual, as compared with the individual's previous behaviour or the behaviour of their peer group. It forms the basis of automated behavioural anomaly detection.

Investigation

All being well, a combination of people and automated systems will detect and report indicators of potential insider risk. However, those indicators will often be tentative and ambiguous. A colleague has been behaving a bit oddly and a few snippets of confidential information have leaked to the media, but are the two related? What is really going on? In some cases, it becomes necessary to investigate the circumstances and establish the facts before deciding what to do next. Many leads will turn out to have innocent explanations, but sometimes they will guide investigators to an active insider. Large organisations may have their own in-house investigative capacity; others call in specialist suppliers. The threshold for launching an investigation is generally quite high, which means that many tentative leads will not be pursued.

At its most basic, an investigation would involve collecting relevant information from corporate systems, such as IT logs or financial records, and interviewing people. More intrusive techniques are available to investigators in high-security government organisations or legally regulated sectors like financial services. Investigators presented with hot leads to a potentially serious insider case would ideally have recourse to inspecting the working patterns, IT network behaviour, and work-related travel of the suspect(s). In high-risk cases, further tools in the investigative toolkit might include OSINT (open-source intelligence), the polygraph, and structured security interviews (see Chapters 5 and 7).

Any organisation will have only a finite capacity to conduct investigations. This capacity should be used to best effect by pointing it at the most promising high-risk leads and ignoring the background noise. Automated systems can assist by collecting, collating, and analysing multiple data streams, picking out the intelligence needles from the data haystack, and presenting them to the human investigators in actionable form.

Sanctions

Organisations should be able to impose reasonable and proportionate sanctions against insiders who are found to have transgressed. Sanctions range from informal warnings, through written warnings, to dismissal and prosecution. The prospect of such sanctions should provide some deterrence, although no one should rely on it.

Sanctions should be applied thoughtfully, because they have the potential to make matters worse if they are perceived to be unfair or unreasonably harsh. A badly handled dismissal could increase the risk if the aggrieved individual decides to take revenge. Unfair or disproportionate disciplinary measures can also affect others, leaving the sanctioned individual's colleagues worrying that they might be next. This could further increase the risk.

Many of those insiders who have broken the law are not prosecuted, either because of a lack of adducible evidence or because their employer prefers to avoid bad publicity. For example, only around three out of five insider fraud cases globally are referred for prosecution.[27] Many of the British and American traitors who spied for the Soviet Union during the Cold War, including all five Cambridge Spies mentioned in Chapter 2, were never prosecuted. Indeed, one of them, Anthony Blunt, was given immunity from prosecution and was later knighted for his services to the Queen.

Exit procedures

Most people who leave an organisation are not active insiders. Even so, departure is a period of heightened risk, and it should be handled with the potential for insider action in mind. People can still cause harm after they have ceased working for an organisation. In-trust measures should therefore encompass the exit process and beyond, to mitigate the risk from so-called bad leavers.

A well-run organisation will unfailingly terminate a leaver's IT access and retrieve their passes. Judgement is required when deciding the best time to do this. Some organisations take a blunt approach: the day that someone is fired or says they are resigning, their access is immediately terminated and they are escorted off the premises. Other organisations with more trusting cultures allow people to work out their notice periods, possibly lasting months, while retaining full access. Both approaches carry risks and benefits. The blunt method could fuel animosity in people who are leaving reluctantly. On the other hand, it is probably the right approach with individuals who have been fired for serious transgressions.

The basic precaution of terminating authorised access offers only limited protection. Inside every leaver's head there will be stored knowledge of the organisation, its methods, products, and people. A determined insider could also take with them data or assets that they have illicitly collected ahead of their departure. There have been numerous cases of bad leavers stealing intellectual property, software, or data shortly before departing, and taking it with them to a new job.

Case histories

2018: An IT consultant hacked an IT consulting company and deleted most of its Microsoft 365 accounts after being sacked. He was jailed.[28]

2016: A former IT engineer for a US law firm was jailed for deleting hundreds of user accounts several months after leaving the company.[29]

2012: An employee of a US energy company pleaded guilty to sabotaging its IT system after discovering he was going to be dismissed. The company's operations were halted for 30 days.[30]

2003: An employee of UK defence company BAE Systems was jailed for attempting to sell defence secrets to a Russian agent when he thought he was going to be made redundant. He was caught in an MI5 sting operation.[31]

Comment

- Insiders may harm their employer shortly before or after leaving.

The exit interview

What else could be done to mitigate the risk from bad leavers? Conducting an exit interview with everyone who leaves is a sensible precaution. Exit interviews are a source of insight into drivers of insider risk, such as interpersonal conflict or toxic managers. They also offer an opportunity to send leavers on their way with slightly warmer feelings towards their former employer. In problematic cases, the exit interview can help to deter a potential bad leaver by reminding them of their contractual obligations and warning them of legal consequences if they transgress. On the other hand, leavers understand that an exit interview is mainly for the benefit of their soon-to-be-former employer, and so may not be forthcoming. When HR asks why they are leaving, they may opt for a palatable explanation like career development, rather than telling the uncomfortable truth that they hate working there.

Discussion points

- Does your organisation do enough about in-trust security?
- Do the leaders of your organisation understand insider risk?
- If you could wave a magic wand, what technology would you invent to support personnel security?
- How has remote working changed insider risk?
- How would an organisation defend itself against AI insiders?
- What are the best indicators of insider risk?

Notes

1 Bunn, M. and Sagan, S. D. (2016). A worst practices guide to insider threats. In *Insider Threats*, ed. by M. Bunn and S. D. Sagan. Ithaca NY: Cornell University Press.

2 Andrew, C. (2009). *The Defence of the Realm: The Authorized History of MI5*. London: Allen Lane.

3 ACFE (Association of Certified Fraud Examiners). (2022). *Occupational Fraud 2022: A Report to the Nations*. https://legacy.acfe.com/report-to-the-nations/2022/

4 Proofpoint. (2022). *The Human Factor 2022. People-Centric Cybersecurity in An Era of User-Based Risks*. www.proofpoint.com.

5 Rice, C. and Searle, R. H. (2022). The enabling role of internal organizational communication in insider threat activity — evidence from a high security organization. *Management Communication Quarterly*, *36*: 467–495.

6 ACFE (Association of Certified Fraud Examiners). (2022). *Occupational Fraud 2022: A Report to the Nations*. https://legacy.acfe.com/report-to-the-nations/2022/

7 Bell, A. J. C., Rogers, M. B., and Pearce, J. M. (2019). The insider threat: Behavioral indicators and factors influencing likelihood of intervention. *International Journal of Critical Infrastructure Protection*, *24*:166–176.

8 Searle, R. and Rice, C. (2018). *Assessing and Mitigating the Impact of Organisational Change on Counterproductive Work Behaviour: An Operational (Dis)trust Based Framework*. CREST. https://eprints.gla.ac.uk/158525/1/158525.pdf; Bell, A. J. C., Rogers, M. B., and Pearce, J. M. (2019). The insider threat: Behavioral indicators and factors influencing likelihood of intervention. *International Journal of Critical Infrastructure Protection*, *24*: 166–176.

9 Agrafiotis, I. et al. (2017). Formalising policies for insider-threat detection: A tripwire grammar. *Journal of Wireless Mobile Networks, Ubiquitous Computing, and Dependable Applications*, *8*: 26–43.

10 See, for example: Alsowail, R. A. and Al-Shehari, T. (2022). Techniques and countermeasures for preventing insider threats. *PeerJ Computer Science*, *8*: e938.

11 See, for example: Yuan, S. and Wu, X. (2021). Deep learning for insider threat detection: Review, challenges and opportunities. *Computers & Security*, *104*: 102221.

12 CERT. (2018). *Common Sense Guide to Mitigating Insider Threats. Sixth edition.* Carnegie Mellon University. https://resources.sei.cmu.edu/asset_files/TechnicalReport/2019_005_001_540647.pdf

13 Greitzer, F. L. (2019). Insider threats: It's the *HUMAN*, stupid! *NW Cybersecurity Symposium*, April 2019. https://doi.org/10.1145/3332448.3332458

14 Theis, M. C. (2020). *Insider Threat or Insider Risk — What Are You Trying to Solve?* Carnegie Mellon University. https://apps.dtic.mil/sti/pdfs/AD1110414.pdf

15 Jeong, M. and Zo, H. (2021). Preventing insider threats to enhance organizational security: The role of opportunity-reducing techniques. *Telematics and Informatics, 63*: 101670.

16 Ball, K. (2021). *Electronic Monitoring and Surveillance in the Workplace. Literature Review and Policy Recommendations.* Publications Office of the European Union, Luxembourg. DOI:10.2760/5137

17 Costa, D. (2020). *Technical Detection Methods for Insider Risk Management.* Carnegie Mellon University. https://apps.dtic.mil/sti/pdfs/AD1110364.pdf

18 Wilder, U. M. (2017). The psychology of espionage and leaking in the digital age. *Studies in Intelligence, 61*: 1–36.

19 Shaw, E. et al. (2013). How often is employee anger an insider risk I? Detecting and measuring negative sentiment versus insider risk in digital communications. *Journal of Digital Forensics, Security and Law, 8*: 39–72.

20 Langer, E. (2022). Ronald Pelton, spy convicted of selling secrets to Soviets, dies at 80. *The Washington Post*, 16 Sept 2022.

21 Luckey, D. et al. (2019). *Assessing Continuous Evaluation Approaches for Insider Threats. How Can the Security Posture of the US Departments and Agencies Be Improved?* RAND National Defense Research Institute. RR2684. Santa Monica CA: RAND Corporation. https://apps.dtic.mil/sti/pdfs/AD1078959.pdf

22 Theis, M. C. (2020). *Insider Threat or Insider Risk — What Are You Trying to Solve?* Carnegie Mellon University. https://apps.dtic.mil/sti/pdfs/AD1110414.pdf

23 See, for example: Legg, P. et al. (2013). Towards a conceptual model and reasoning structure for insider threat detection. *Journal of Wireless Mobile Networks, Ubiquitous Computing, and Dependable Applications, 4*: 20–37; Nurse, J. R. C. et al. (2014). Understanding insider threat: A framework for characterising attacks. *2014 IEEE Security and Privacy Workshops.* https://doi.org/10.1109/SPW.2014.38; Elifoglu, I. H., Abel, I., and Taşşeven, Ö. (2018). Minimizing insider threat risk with behavioral monitoring. *Review of Business: Interdisciplinary Journal on Risk and Society, 38*: 61–73; Homoliak, I. et al. (2018). Insight into insiders and IT: A survey of insider threat taxonomies, analysis, modeling, and countermeasures. *ACM Computing Surveys, 52*: 1–40; Georgiadou, A., Mouzakitis, S., and Askounis, D. (2021). Detecting insider threat via a cyber-security culture framework. *Journal of Computer Information Systems, 62*: 706–716.

24 Cappelli, D., Moore, A., and Trzeciak, R. (2012). *The CERT Guide to Insider Threats.* Carnegie Mellon University. www.asecib.ase.ro/cc/carti/The%20CERT%20Guide%20to%20Insider%20Threats%20%5b2012%5d.pdf

25 Greitzer, F. L. et al. (2019). Design and implementation of a comprehensive insider threat ontology. *Procedia Computer Science, 153*: 361–369.

26 Greitzer, F. L. (2019). Insider threats: It's the *HUMAN*, stupid! *NW Cybersecurity Symposium*, April 2019. https://doi.org/10.1145/3332448.3332458

27 ACFE (Association of Certified Fraud Examiners). (2022). *Occupational Fraud 2022: A Report to the Nations.* https://legacy.acfe.com/report-to-the-nations/2022/

28 CDSE. (2023). www.cdse.edu/Portals/124/Documents/casestudies/case-study-deepanshu-kasaba-kher.pdf

29 DoJ. (2016). www.justice.gov/usao-ndtx/pr/former-law-firm-it-engineer-convicted-computer-intrusion-case-sentenced-115-months

30 DoJ. (2014). www.justice.gov/usao-sdwv/pr/former-network-engineer-pleads-guilty-crashing-employers-computer-system

31 Dodd, V. (2003). Would-be spy gets 10 years. *The Guardian*, 5 Apr 2003.

9 Foundations

Reader's Guide: This chapter describes the cross-cutting functions needed to underpin pre-trust and in-trust measures, including governance, ethics, culture, and risk management.

Pre-trust and in-trust personnel security measures require solid foundations to build a coherent personnel security system. The capabilities and attributes that provide those foundations include:

- Governance
- Ethics
- Leadership
- Management
- Deterrence communication
- Risk management
- Incident and crisis management
- Security culture
- Asset management
- Information sharing
- Assurance

For pre-trust and in-trust security measures to work well together, they must be supported by good governance, ethical leadership, competent management, dynamic risk assessment, systematic information sharing, independent assurance, and a positive security culture, among other things. These foundational capabilities, in combination with the pre-trust and in-trust measures, form the threefold model of personnel security depicted in Figure 6.2.

Governance

The first question to ask when forming an impression of any organisation's security is: 'Who's in charge?' To put it another way: What are the governance arrangements for security?

Governance is the system by which an organisation is directed, managed, and controlled. Good governance is a prerequisite for good protective security. It has three core components:

- *Accountability*: Who is ultimately accountable for security risks?
- *Responsibility*: Who is responsible for doing the work to understand and manage the risks?
- *Authority*: Who has the authority and the requisite resources to make decisions and take the necessary actions?

DOI: 10.4324/9781003329022-12

A lack of clarity about governance is widely associated with substandard security and unmitigated risk. Indeed, poor governance is a root cause of organisational problems and poor performance in other domains besides security.

Poor governance is unfortunately the norm with personnel security. In many large organisations, responsibility for dealing with insider risk is divided between functions like HR, security, legal, internal audit, compliance, and cyber security, with tenuous arrangements for sharing relevant information. In small organisations, the responsibility often sits uneasily in HR, if anywhere at all. The chances are that no one at board level is accountable for insider risk, although it is a racing certainty that a named executive will be accountable for cyber risk. Most large organisations have a Chief Information Security Officer (CISO) but why is there no such thing as a Chief Personnel Security Officer?

Poor governance leaves no one feeling accountable for the risk or empowered to make difficult decisions. When no one feels accountable or responsible, warning signs will go undetected and overlooked. As we have seen, a recurrent feature of insider cases is the hindsight recognition of red flags which were ignored until it was too late. Poor governance is a recipe for gaps, overlaps, and inefficiency, as people stick within their silos and assume someone else is tackling the problem.

Good governance means making a named person at the top of the organisation accountable for insider risk; spelling out in clear and unambiguous terms who is responsible for performing the personnel security functions; and giving those people the authority and the resources they need to do their jobs.

Non-executive directors and independent advisers have a role to play by calling leaders to account. Table 9.1 lists some questions they could ask leaders if they suspect that personnel security is being undercooked.

A standard governance model for risk management in larger organisations is known as the three lines of defence.[1] The first line of defence is the operational management – namely, the people responsible for operating the functions on a day-to-day basis. The second line consists of central functions that oversee or specialise in risk management across the organisation. They monitor risk management processes and support front-line managers with specialist advice. The

Table 9.1 Questions for leaders

- Do we have a solid understanding of what insider risk means for this organisation, including the potential impact of a major insider attack?
- Is our investment in personnel security proportionate to the risk, bearing in mind the potential losses that could result from a major insider attack?
- Do we recognise insider risk as a distinct risk in its own right (as opposed to a cross-cutting factor that facilitates cyber security risk)?
- How do we judge the severity of our insider risk and the effectiveness of our personnel security defences? Are we seeing the information needed to form this judgement?
- Who is accountable for insider risk at the most senior level?
- Who is responsible for understanding and managing insider risk? Do they have the necessary authority and resources?
- Are the different functions that contribute to personnel security sharing information and collaborating?
- What are our most important assets? Where are they and who has access to them?
- How might our most important assets be compromised by insider action?
- Is our approach to security holistic?
- Do we understand the insider risk in our supply chain?

third line of defence is an assurance function which is generally provided by an internal audit team. Their role is to give the organisation's leaders independent and objective advice on how well the risk management functions are working.

Ethics

Strong ethical values are a bedrock of good personnel security and organisational success. Organisations that lose sight of ethical values are inviting disaster, and not just because their security will be flawed. Unethical attitudes enable bad decisions that eventually cause problems. Well-known examples include the collapse of the US company Enron, Volkswagen's 'dieselgate' scandal, and the furore over MPs' expenses that engulfed the UK Parliament in 2009 and caused lasting damage to its reputation. The commission of inquiry into the global financial crisis of 2007–2008 concluded that one of its root causes was a systemic breakdown in ethics.[2] Ethical violations can cause severe reputational damage. They should feature in any organisation's register of risks.

Societies cannot rely on legislation alone to prevent bad behaviour. Just because something is lawful does not mean it is the right thing to do. Similarly, organisations cannot rely on policies and compliance procedures alone to ensure that people do the right thing. Ethical values provide the foundations on which policies and rules should rest, and an environment in which good behaviour, including good security behaviour, can flourish. Research has confirmed that employees are more likely to trust their organisation if it upholds high ethical standards.[3]

Ethical principles are crucial in guiding decisions during times of crisis or rapid change, when prescriptive rules are not enough. Even so, they cannot always compensate for the effects of inadequate resources or dysfunctional working practices, when pressing demands to get the job done may put ethical values under strain. The stronger the values, the less likely they are to buckle.

What sorts of ethical values or standards should an organisation adopt? There are many published frameworks to choose from. However, a good starting point would be the seven Principles of Public Life that were first set out in 1995 by the UK's Committee on Standards in Public Life, and which apply to everyone in public service, including civil servants and government ministers.[4] They are:

- Selflessness
- Integrity
- Objectivity
- Accountability
- Openness
- Honesty
- Leadership

The seven principles have stood the test of time and, although designed originally for public servants, they have a universal relevance. More simply, individuals who find themselves confronting an ethical dilemma could do worse than ask themselves a few questions, such as: How would I feel if this happened to me? How would I feel explaining it to my family and friends? And what would happen if everyone did it?

Ethical cultures do not arise spontaneously: they require sustained effort by leaders.[5] Setting, communicating, and demonstrating ethical values is a core responsibility of leaders at all levels. They must be clear and consistent about how they expect their workforce to behave. They must also demonstrate that behaving unethically will have consequences. Ethical values should be considered when recruiting new people (see Chapter 7). If a person does not share the ethical values of the organisation, it might be unwise to recruit them.

Leadership

An organisation is unlikely to enjoy strong personnel security unless its leaders understand insider risk and devote some attention to it. Leaders need to ensure that, somewhere in the organisation, insider risks are being identified, assessed, and properly managed, and that there are clear governance arrangements for doing so. They should take a strategic view of insider risk in the broader context of other security and corporate risks.

As well as giving personnel security the attention it deserves, leaders have a direct influence on insider risk itself. Workers who witness their leaders ignoring the rules may feel entitled to follow suit. Leaders who behave unethically or tolerate injustice will generate more insider risk. Conspicuous discrepancies between a corporate rhetoric of high-minded values and a daily reality of unprincipled behaviour will create breeding conditions for insider risk. Bad leaders make bad cultures, which give rise to bad security behaviours. To give just one example, a study of a high-security organisation found that immoral leaders undermined trust and increased security risks. Rule-breaking by leaders was associated with counterproductive work behaviours, which are potential precursors to insider risk.[6] There is much truth in the adage that a fish rots from the head down.

By the same token, leaders who set a positive example and instil a culture of integrity and trust can do much to mitigate insider risk, as well as making their organisation more productive and resilient. Ethical leadership builds confidence and mutual trust, leaving people to focus on doing their jobs rather than watching their backs. Trusted leaders are particularly important during turbulent periods like economic down-turns, major reorganisations, or security crises.

To be trusted, leaders must demonstrate the core components of trustworthiness – namely, benign intentions, integrity, competence, and consistency. Of these, competence is usually the most visible and easiest to judge. However, trust in leaders is more than just a product of their personal qualities. Leaders are also judged by the culture of their organisation. A leader who exudes benign intentions and energetic competence may still not be trusted if they preside over an unethical culture, or if they manifestly do not trust their workforce. Trust should be reciprocal.

Management

Competent management is a crucial, if unglamorous, bulwark against insider risk. As noted before, the best detectors of early warning signs are the potential insiders' managers and colleagues. The ability of managers to fulfil this function has been made more complicated (but by no means impossible) by the move to remote working.

Conversely, bad managers are a common cause of the disgruntlement that fosters insider risk. Disaffected employees are often found in a cluster with a bad manager as the common factor, prompting the wise observation that rotten barrels make rotten apples. Toxic working environments created by bad managers are fertile habitats for insiders. They also make insiders harder to detect, because people are less likely to notice and report concerns.

The HR function has an integral role in personnel security. It can advise managers on how to recruit the right people, handle disputes, deal with grievances, train and develop people, manage departures, and avoid inadvertently breaching employment law. Personnel security specialists should make HR specialists their best friends.

The wrong kind of HR can be part of the problem. In some organisations, the HR function expands beyond its core role. Instead of enabling the organisation to fulfil its mission, HR becomes an end in itself, as though demonstrating HR best practice had become the mission. Cumbersome HR processes consume people's time, managers have less autonomy to manage, and risk-averse leaders unquestioningly follow HR directions for fear of complaints or litigation. In one study, researchers found that heavy-handed and inconsistent behaviour by HR had created an unhealthy culture in which employees were reluctant to report concerns about potential insider risk for fear of triggering excessive disciplinary responses. HR had eroded the confidence of managers to manage personnel problems, leaving gaps where insider risk could ferment.[7]

Efficiency pressures have driven some organisations towards HR practices that may not be conducive to building trust. Internal job markets and self-service HR systems might be convenient (at least, for the people who operate them) but they can also be impersonal and alienating. In the far distant past, when HR was called Personnel, the processes were often more personal, involving face-to-face interactions between humans. As is often the case, changes that were intended to improve matters have had unintended consequences.

Deterrence communication

We saw in Chapter 1 that the only three ways to reduce any security risk are by reducing threat, reducing vulnerability, or reducing impact (or some combination of the three). The purpose of deterrence is to reduce security risk by influencing the intentions of threat actors and hence reducing threat. In the case of personnel security, this means influencing the potential insiders and the external threat actors who might try to cultivate and direct them. If the probability of detecting an insider is much less than one – which it generally is – then the deterrence arising from the prospect of being detected could, in theory, be more effective in reducing insider risk than the detection process itself.

The existence of pre-trust personnel security measures should act as a mild deterrent which discourages some would-be insiders from applying. An individual with pre-existing malign intentions might hesitate at the prospect of having their official records checked or their history probed in an interview. Deterrence is hard to quantify, however, because the potential insiders who are deterred from applying rarely publicise the fact. Nonetheless, it is a free win because the pre-trust measures are there anyway.

A more active approach involves the use of deterrence communication to influence potential insiders, both during the pre-trust recruitment phase and thereafter. Deterrence communication is intended to convey a sense that the organisation has effective security and therefore any would-be insider will be found out. The three main audiences for deterrence communication are potential insiders who are thinking of applying (pre-trust), potential insiders who are already part of the workforce (in-trust), and external threat actors who might try to recruit an insider.

Subtle ambiguity is a feature of good deterrence communication. It paints a picture of dynamic and multi-layered defences without revealing details of exactly how they work. The aim is to unsettle the potential insider by creating a sense of jeopardy, while simultaneously reassuring everyone else that the organisation is looking after its people and assets. For instance, a recruitment website might refer to there being a thorough pre-employment screening process, without specifying precisely how this is done. A desirable feature of deterrence communication – and indeed *all* communication – is active engagement with the target audience. The aim of communication should be to change people's attitudes and behaviour, not just spray them with information, and the best way of doing that is by engaging their interest and, where possible, enabling their participation.

A more vigorous technique involves deliberately sowing distrust between would-be insiders and the external threat actors with whom they might collaborate. Hostile foreign intelligence services, organised crime groups, terrorist organisations, and other external threat actors are implicitly encouraged to suspect that any insider who contacts them might be an undercover police or intelligence officer. Similarly, would-be insiders are led to suspect that if they were to volunteer their services, they might end up talking to an undercover officer from their own side. Publicity about sting operations should have a chilling effect on both hostile parties. The mutual suspicion might make them more reluctant to initiate an approach and more likely to decline if they are approached.[8]

Case history

2012: Royal Navy Petty Officer Edward Devenney was jailed for collecting classified information with the intention of passing it to Russia. He was caught in a sting operation in which he offered nuclear submarine secrets to undercover MI5 officers posing as Russian intelligence officers.[9]

Comment

- Media publicity about successful sting operations might give other potential insiders pause for thought.

Risk management

Risk management is an essential element of any security regime. The purpose of personnel security, as set out in Chapter 6, is to understand and manage the security risk arising from insiders. Understanding the risk entails developing an awareness of the current threats, vulnerabilities, and impacts that constitute the risk. Managing the risk involves deciding how much risk to tolerate, and then acting to mitigate the risk if it exceeds a tolerable level. A simple representation of security risk management is the three-step model shown in Figure 9.1.

Security risks are dynamic and adaptive, which is why the three steps must be repeated continuously. Let us consider each of the steps in more detail.

Understand

Before a security risk can be effectively managed it must be understood. Lack of understanding is itself an indicator of heightened risk. Security risks are always evolving, so understanding must be kept current.

The process of developing and maintaining a current understanding of insider risk (and other security risks) has two main elements:

- *Risk discovery*: finding out what is going on by gathering information
- *Risk assessment*: making sense of the information and forming a view of the risk

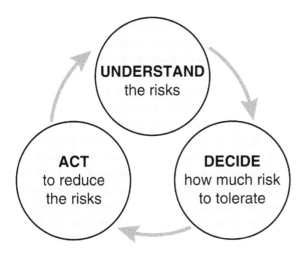

Figure 9.1 A simple three-step model of security risk management

Source: After Martin, 2019.

The discovery element involves collecting and collating relevant information. The more the merrier, within reason. Risk discovery should encompass threat intelligence about the identities, intentions, and capabilities of external threat actors and potential insiders; vulnerability analysis to identify gaps or weaknesses in security; and impact analysis to consider the likely consequences of an attack. The risks should be clearly defined.

Risk discovery should extend its time horizon beyond the present to identify new and emerging risks and contemplate what the future might bring. A conventional way of doing this is with horizon scanning, which involves systematically exploring a wide variety of information sources to identify potential new risks, and considering what might happen if they were to materialise. Horizon scanning is a creative process designed to pick up early warning signals and explore possible futures. It is not about attempting to predict the future. When scanning the risk horizon, it is best to avoid three common pitfalls: ignoring weak or uncertain signals; paying insufficient attention to complex risks that are difficult to understand; and relying too much on extrapolating from current trends. From time to time, weak signals will presage large and complex risks that diverge dramatically from current trends.

Risk assessment involves bringing together the threat, vulnerability, and impact components of risk to form an evidence-based understanding of its nature and severity. It should include a consideration of internal and external factors that might influence insider risk, such as organisational change, employment disputes, the political and economic environment, and so on. Risks should be identified and assessed regardless of how far the organisation feels able to control them. A risk is a risk. Whether you can make it go away is another matter.

Decide

Having developed a sufficient understanding of the risk, the next step is to decide whether it is tolerable. If the answer is 'no', as it often is, then the third step is to act to mitigate the risk and, if possible, reduce it to a level that *is* tolerable.

Judging whether complex security risks are tolerable is more art than science, because there are few objective metrics or criteria by which to make the judgement (see Chapter 10). The decision will often come down to a professional judgement about the feasibility, affordability, and acceptability of the additional measures that would be required to reduce the risk.

Language is important here. Specialists in risk management often refer to an organisation's risk appetite, by which they mean the amount of risk the organisation is willing to accept in pursuit of some goal. The same specialists often advocate the use of quantitative metrics of risk appetite. The concept of risk appetite can be helpful when applied to the types of risks for which it was originally invented – notably, project and programme risks, conventional safety risks, engineering risks, insurance risks, and some kinds of financial and business risks. For instance, a bank might decide to launch a new financial product if its actuarial modelling predicts that the expected profit will significantly exceed its risk appetite, as measured by the predicted loss through fraud. However, this sort of approach is unhelpful when applied to security risks.

Security risks are different, for reasons discussed previously. They are dynamic and adaptive. The most serious security risks materialise infrequently and each one is unique, so there is little actuarial data on which to base quantitative predictions about likelihoods or impacts. Furthermore, security risks are not elective: they are imposed on us by threat actors. An organisation might choose to accept the risk associated with launching a new product, but it does not choose to be threatened by insiders or hostile foreign states. We do not have an 'appetite' for security risks: we *tolerate* them.

Governments notably avoid the language of risk appetite when talking about terrorism. No government would dream of announcing that its risk appetite for terrorism was, say, five fatal attacks a year, even if such a statement made sense in purely analytical terms (which it does not). When confronting the most serious risks to national security, like terrorism and hostile foreign state activity, governments follow an implicit strategy of seeking to reduce the risk to the lowest level possible within the constraints of feasibility, affordability, and acceptability. A similar approach to risk tolerance is taken in the civil nuclear industry, where the explicit aim is to reduce safety and security risks to a level that is ALARP (As Low As Reasonably Practicable).

Act

If an organisation decides that a security risk exceeds its tolerance, it should act to mitigate the risk. For insider risk, that means bolstering personnel security defences in ways that reduce some combination of threat, vulnerability, and impact.

A common weakness in risk management is the tendency to jump straight to action, shortcutting the crucial process of understanding. This problem is epitomised by the misuse of risk registers. A security risk register is an inventory of specific security risks, together with the corresponding actions to mitigate them. The risk register will also specify who is responsible for implementing each action and the timeline for doing so. There is nothing intrinsically wrong with risk registers, provided they genuinely reflect the current understanding of the ever-changing risks. However, risks registers are often compiled without sufficient time being devoted to understanding the risks, and without a rigorous process for regularly reviewing and updating the risks as they evolve. The result can be, at worst, a fossilised list of increasingly irrelevant risks which do not reflect the current risk picture, but which nonetheless generate a lot of superficially reassuring activity.

Good risk management also requires effective mechanisms for reporting and escalating risks. Senior managers, leaders, and the board of an organisation should be kept informed about security risks through reporting that is timely, accurate, actionable and – crucially – understandable, so that they can make well-informed decisions. Risk reporting diminishes in value if the processes are over-engineered and the content becomes mired in detail. Risk reports are there to help decision makers think clearly, not as a showcase for the methodological prowess of the specialists who produce them. There should also be a robust escalation mechanism for drawing leaders' attention to new or unusual risks.

Incident and crisis management

If prevention fails, and a risk materialises, the only remaining option is to confront the problem and try to stop it getting worse. This requires a capability to manage incidents and crises. In the context of insider risk, it means having a personnel security system that includes functions for identifying insider incidents and managing their consequences.

Most organisations have some capabilities for coping with security incidents and crises. (If not, they soon create them after a disaster.) These capabilities include incident management, crisis management, business continuity planning, and disaster recovery. They mitigate the risk by helping to reduce the impact when something bad happens. However, they are mainly designed to cope with security risks in the cyber and physical domains, such as ransomware attacks or terrorism. Fewer organisations have specific capabilities for managing security incidents arising from insiders.

What would an insider incident management capability look like? The starting point would be an ability to recognise that an insider incident was underway or had happened. This might not be obvious. A 'low and slow' attack by a capable covert insider might produce few noticeable signs of anything untoward. Any signs that do appear might not be detected until sometime later – for example, when stolen confidential information appears in the media, or an adversary exploits the knowledge they have purloined through espionage. In some cases, however, it will be apparent that an insider is active. At that point it helps to have an insider incident management plan.

Insider incidents are more likely to be handled well if the organisation has a clear definition of what constitutes an incident, as distinct from the normal rough-and-tumble of everyday security business. Different organisations apply different thresholds beyond which an everyday event becomes an incident, an incident becomes a crisis, and a crisis turns into a disaster. According to one guideline, a crisis is an acute period of existential risk, while a disaster is what happens if the risk materialises.

Insider incidents are valuable sources of data about insider risk. It is a good idea to collect that data and compile a database of incidents. Some organisations do not recognise that they already possess relevant data, such as records of theft, fraud, investigations, and security violations. The database should ideally be extended to include data about near-miss events that fall short of being classed as insider incidents, plus other data that might help to understand risk factors and trigger events. The resulting pool of data can be analysed for insights into the origins of insider incidents and the lessons learned. Organisations can further improve their understanding by sharing their insights with other organisations.[10]

Security culture

Security culture is sometimes spoken of as though it meant something faintly magical. We are told that culture eats strategy for breakfast, and that security culture is the key to better security. Organisations have increasingly invested in programmes to improve their corporate or security cultures. But what does 'security culture' mean and how can we measure it? What is the difference between culture and *security* culture? What does a good security culture look like and how can we build one? These are questions to which the answers are sometimes sketchy.

Security culture is a vague term. One of the simplest definitions is that an organisation's security culture is its consistent way of behaving towards security, or 'the way we do security around here'. According to another version, an organisation's culture is analogous to an individual's personality: it describes their consistent tendency to behave in certain ways under certain conditions. Accordingly, an organisation's *security* culture is analogous to its personality with regard to security. Either way, 'security culture' is essentially a label for a collection of security-related behaviours. The consensus view that security culture has a strong influence on security behaviour is therefore unsurprising, as they are basically two ways of saying the same thing. It is generally more fruitful to focus on specific behaviours than the catch-all of culture.

An organisation's culture can be a barrier to good personnel security. People are less inclined to do the right thing if they are surrounded by colleagues and leaders who get away with breaking the rules. Leaders can inadvertently make things worse, even when trying to do the right thing. Relentless positivity about security culture can do more harm than good if it encourages a complacent attitude of Not In My Organisation (NIMO).[11] Even in high-trust organisations, there is always a finite risk that someone will behave badly.

Compliance culture versus concordance culture

Building a good security culture is difficult without first agreeing what good would look like. At its simplest, a good security culture is one in which people consistently do the right things regarding security. They understand and follow the rules and report any concerns, including possible signs of insider behaviour. However, cultures differ widely between organisations, industries, and nations, and there is no universal template.

A broad distinction may be drawn between two different styles of (security) cultures:

- *Compliance cultures*, in which the emphasis is on checking and enforcing people's adherence to prescriptive rules by means of management oversight and auditing.
- *Concordance cultures*, in which people understand the risks, actively want to do the right thing, and are helped to do so.[12]

Concordance cultures tend to be more successful in producing sustained improvements in security behaviour. Organisations with a concordance culture actively promote risk awareness and high ethical standards. They provide positive incentives for good security behaviour. By contrast, the worst kinds of compliance cultures rely on negative incentives and a box-ticking mentality in which security behaviour is more performative than effective. Crude compliance cultures ignore the reality that people will fail to comply with security rules for all sorts of reasons, such as being pressured to complete a task, not understanding the rules, or thinking the rules are pointless.

Compliance cultures can work better in regulated industries and organisations that have a strong focus on safety, where a strict rules-based approach may be more appropriate. However, what works for safety may not work equally well for personnel security, where the risk emanates from covert adversaries who will find ways of breaking the rules without being caught.

Security culture or just culture?

The distinction between organisational culture and security culture can be unclear and unhelpful. An organisation's security culture is a sub-set of its organisational culture, not an independent entity. Building a positive security culture would be difficult without a positive organisational culture to support it.

Organisations with flawed cultures will struggle to build good security cultures. An example of the read-across from culture to security culture is that of UK policing. In recent years, a number of serving police officers have been convicted of serious criminal offences or dismissed for gross misconduct. These events have involved only a small minority of officers. Nonetheless, they have harmed the police service by damaging public trust and confidence. They have also highlighted deficiencies in police personnel security and culture. The concerted efforts that are being made to address these problems are rightly directed at the culture of these organisations, not just their security culture.

An independent review of the UK's largest police force, the Metropolitan Police Service, found that it had lost the trust of the public.[13] The review's final report, published in 2023, concluded that the force had routinely put its own interests above the needs of the public it was meant to serve, and could no longer be trusted to fix its 'systemic and fundamental' issues. Inadequate vetting and systemic failures to tackle misconduct had allowed predators to remain within its ranks. The abduction, rape, and murder of Sarah Everard by a serving officer in 2021 should have led to fundamental reform. Instead, the force had responded with complacency and defensiveness and 'preferred to pretend that their own perpetrators of unconscionable crimes were just "bad apples"'.[14] It is hard to see how a positive security culture could thrive in any organisation without a healthy organisational culture.

Case histories

2022: A former Thames Valley police officer was jailed for sexual offences after pursuing several women whom he had met while on duty. He was described as a sexual predator. He resigned before the trial.[15]

2022: A Metropolitan Police officer was jailed for child sex offences after grooming a 13-year-old girl he met online and travelling to meet her. When arrested, he was found to be carrying condoms, lubricant, and a Viagra-like drug. The person he had been grooming turned out to be an undercover police officer.[16]

2021: Two Metropolitan Police officers were jailed for taking and sharing photos of two female murder victims while they were guarding the crime scene.[17]

Comments

- Cases like these damaged public trust in the police, undermining their ability to perform their vital functions.
- They imply a problem with police culture, as well as its personnel security processes.

Asset management

Controlling people's access to assets is an important element of in-trust personnel security, as discussed in Chapter 8. Before an organisation can apply access controls, it must understand the assets it wants to protect. This is easier said than done. Some key assets are obvious: a government agency's classified information, a bank's money, a technology company's intellectual property, a museum's precious exhibits, and so on. But not all organisations give sufficient thought to their less obvious or less tangible assets, like personal data, reputation, know-how, people, or infrastructure, and they may not fully appreciate the harm they would suffer if these assets were lost or damaged. The best protective security systems include processes for identifying, defining, and managing the organisation's virtual and physical assets.

Classifying assets

Having located and identified the assets to be protected, a desirable further step is to classify those assets according to their value – more precisely, the harm that would be incurred if they were stolen, damaged, or otherwise compromised. The simplest possible classification scheme would divide assets into two categories: the most valuable (crown jewels) and the rest. A more graduated approach with three or more levels would allow finer distinctions, albeit at the cost of more complication and administrative overhead.

Well-known examples of asset classification are the marking schemes used by governments to classify official information. For example, the UK government's current classification scheme distinguishes between three main levels (Official, Secret, and Top Secret) and adds various codewords to specify other characteristics and handling restrictions. A simple three-level classification scheme for an organisation might similarly divide its information assets into three types: Green (least sensitive; open to anyone in the organisation); Amber (for distribution only within the recipients' teams or business areas); and Red (the most sensitive information; for sharing only with named addressees).

Once an organisation has identified and classified its assets, it can then apply a graduated and risk-based approach to its protective security, with stronger protection for the more valuable assets. The alternative is a blanket approach in which everything is protected to the same extent, which in practice often means under-protecting some assets. An additional benefit of classification is helping to raise awareness by reminding people of the value of the assets to which they have access. People can become blasé if they are used to dealing with valuable assets every working day.

Information sharing

Personnel security depends on the regular sharing of relevant information between different corporate functions such as security, HR, IT, legal, audit, and compliance. Much of the information needed to detect the early warning signs of insider risk will sit in the HR and cyber security functions, but organisational silos and anxieties about confidentiality may get in the way of sharing. Cooperation between functions is vital for good personnel security.

Information relevant to insider risk could come from many different sources across an organisation. These are likely to include: confidential reporting channels; investigations; HR personnel, disciplinary, and casework records; IT authentication, access, and printing logs; data loss prevention logs; mobile device management logs; automated monitoring logs; travel records; physical security access and intruder detection logs; performance appraisals; expenses records; emails; phone records; internal audit reports; declarations of conflicts of interest; legal advisory cases; and staff opinion surveys, to name a few.[18]

Bringing all these data and information streams together in one place and making sense of them is a non-trivial undertaking that few organisations attempt. Rather than trying to do it by re-engineering the organisation's management structures, a more feasible approach is to convene an informal insider risk working group consisting of representatives from the most relevant functions. This should create opportunities to share information across silos and spread greater understanding of the risks. A more structured approach would give the working group the authority to commission investigations and coordinate the handling of insider cases.

Assurance

Assurance means actively verifying that all is as it should be. In the context of insider risk, assurance refers to processes for checking that the expected personnel security defences are in place and working as intended. An organisation might believe it has strong defences, but discover through assurance that the reality is less impressive. For instance, its pre-employment screening might turn out to be less effective at detecting fraudulent applicants than was supposed.

The main ways of obtaining assurance about personnel security are through testing and internal audits. Further assurance can be provided by external bodies like independent reviewers, external auditors, and accreditors. Assurance must be independent and objective. Those who operate and manage the security functions should not be the only ones to provide assurance about their effectiveness, which would amount to marking their own homework.

Management oversight and audit can give some assurance that security is functioning as intended. However, the best way of discovering how well it really works is to test it. Testing can take various forms, ranging from desktop exercises to red teaming, in which role players simulating threat actors attempt to defeat the security. Red teaming can give defenders actionable insights into how real threat actors could exploit vulnerabilities. Testing has much to recommend it. In addition to finding out how well the security works, it highlights areas for improvement and enables the participants to practise their skills.

A substantial proportion of an organisation's insider risk will sit in its supply chain. Outsourced security personnel may have wide-ranging access to its buildings and facilities, while outsourced IT providers may have administration rights on its corporate systems. Assurance should therefore extend beyond in-house personnel to cover the supply chain.

Discussion points

- Who is in charge of personnel security in your organisation?
- Who is accountable for insider risk in your organisation?
- Do you trust your leaders?
- Is your organisation ethical?
- How can we improve the communication and understanding of insider risk?
- How could potential insiders be deterred or disrupted before they cause harm?
- How would you prepare an organisation for dealing with a major insider incident?
- What would be the job description for a Chief Personnel Security Officer?

Notes

1 HMG. (2020). *The Orange Book: Management of Risk — Principles and Concepts.* https://assets.pub
lishing.service.gov.uk/government/uploads/system/uploads/attachment_data/file/866117/6.6266_HM
T_Orange_Book_Update_v6_WEB.PDF
2 Botsman, R. (2017). *Who Can You Trust? How Technology Brought Us Together — And Why It Could Drive Us Apart.* London: Portfolio Penguin.

 3 Searle, R. et al. (2011). Trust in the employer: The role of high-involvement work practices and procedural justice in European organizations. *International Journal of Human Resource Management*, *22*: 1069–1092.
 4 CSPL. (2021). *Upholding Standards in Public Life*. November 2021. https://assets.publishing.service. gov.uk/government/uploads/system/uploads/attachment_data/file/1029944/Upholding_Standards_in_ Public_Life_-_Web_Accessible.pdf
 5 CSPL. (2023). *Leading in Practice*. January 2023. https://assets.publishing.service.gov.uk/governm ent/uploads/system/uploads/attachment_data/file/1130992/CSPL_Leading_in_Practice.pdf
 6 Searle, R. H. (2022). Trust = confidence + vulnerability: The role of the leader. *CREST Security Review*, *14*: 18–19. crestresearch.ac.uk.
 7 Rice, C. and Searle, R. H. (2022). The enabling role of internal organizational communication in insider threat activity — evidence from a high security organization. *Management Communication Quarterly*, *36*: 467–495.
 8 Hegghammer, T. and Dæhli, A. H. (2016). Insiders and outsiders. A survey of terrorist threats to nuclear facilities. In *Insider Threats*, ed. by M. Bunn and S. D. Sagan. Ithaca NY: Cornell University Press.
 9 BBC. (2012). www.bbc.co.uk/news/uk-20701842
10 Miller, S. (2020). *Leveraging Insider Threat Incident Data and Information Sharing for Increased Organizational Resiliency*. Carnegie Mellon University. https://apps.dtic.mil/sti/pdfs/AD1110241.pdf
11 Bunn, M. and Sagan, S. D. (2016). A worst practices guide to insider threats. In *Insider Threats*, ed. by M. Bunn and S. D. Sagan. Ithaca NY: Cornell University Press.
12 Martin, P. (2019). *The Rules of Security: Staying Safe in a Risky World*. Oxford: Oxford University Press.
13 Casey, B. (2023). *An Independent Review into the Standards of Behaviour and Internal Culture of the Metropolitan Police Service*. Baroness Casey Review. Final Report. March 2023. www.met.police.uk/ SysSiteAssets/media/downloads/met/about-us/baroness-casey-review/update-march-2023/baroness-casey-review-march-2023a.pdf
14 Ibid.
15 BBC. (2022). www.bbc.co.uk/news/uk-england-berkshire-61269842
16 CPS. (2022). www.cps.gov.uk/thames-and-chiltern/news/metropolitan-police-officer-francois-olwage-jailed-child-sex-offences
17 BBC. (2022). www.bbc.co.uk/news/uk-england-london-62095812
18 Costa, D. (2020). *Technical Detection Methods for Insider Risk Management*. Carnegie Mellon University. https://apps.dtic.mil/sti/pdfs/AD1110364.pdf

10   Models and metrics

Reader's Guide: This chapter describes ways of assessing insider risk and personnel security.

Anyone attempting to grapple with insider risk will need answers to three big questions:

1. *What problem are we trying to solve?* What is the nature and scale of the insider risk we are facing?
2. *What tools do we need to solve the problem?* What kind of personnel security system is required?
3. *How well are we doing?* How effective is our personnel security?

Models and metrics are tools for capturing, assessing, and communicating the information needed to answer these questions. A model is a way of representing the essential components of a personnel security system. A metric is a specified variable that is measured in some way. Metrics could be used to measure insider risk or the effectiveness of personnel security defences. Metrics are not just numbers derived from counting things that can be counted. Their purpose is to convey meaningful information in ways that influence thinking and decision making. Some of the most informative metrics in protective security are qualitative.

Models of personnel security

A personnel security model is a simplified way of depicting the building blocks that are required to construct a functional personnel security system. Organisations can compare their own personnel security defences against an authoritative external model to help them judge whether theirs are up to the mark. For those who want to make wide-ranging improvements, a model can provide the design basis for an insider risk programme.

The simple threefold model

We saw in Chapter 6 how a personnel security system could be represented by a simple threefold model comprising pre-trust measures, in-trust measures, and foundations, as shown in Figure 6.2. The main types of pre-trust measures, in-trust measures, and foundations were described in Chapters 7, 8, and 9, respectively, and they are summarised here in Table 10.1.

The Personnel Security Maturity Model (PSMM)

A model that is widely utilised by security professionals is the Personnel Security Maturity Model (PSMM) produced by the National Protective Security Authority (NPSA, formerly

DOI: 10.4324/9781003329022-13

Table 10.1 Components of a threefold model of personnel security

PRE-TRUST MEASURES	IN-TRUST MEASURES
Interviews | Record checking | Open-source intelligence | Psychometric testing	Access controls | Exit controls | Behavioural controls | Awareness raising, training, and communication | Reporting channels and management oversight | Automated monitoring | Investigation | Sanctions | Exit procedures

FOUNDATIONS

Governance | Ethics | Leadership | Management | Deterrence communication | Risk management | Incident and crisis management | Security culture | Asset management | Information sharing | Assurance

CPNI), the UK government's national technical authority for personnel and physical protective security.[1]

As described in Chapter 6, the PSMM consists of seven 'pillars', each representing a critical component of personnel security. They are: Governance and leadership; Insider risk assessment; Pre-employment screening; Ongoing personnel security; Monitoring and assessment of employees; Investigation and disciplinary practices; and Security culture and behaviour change. The seven pillars of the PSMM correspond with the simple threefold model in Table 10.1 as follows:

1. *Pre-trust measures*: Pre-employment screening
2. *In-trust measures*: Ongoing personnel security; Monitoring and assessment of employees; Investigation and disciplinary practices
3. *Foundations*: Governance and leadership; Insider risk assessment; Security culture and behaviour change

The PSMM includes an assessment process by which an organisation can judge its overall level of personnel security maturity – that is, the extent to which the key elements of the model are present and functioning effectively. The maturity assessment is essentially subjective, relying on structured professional judgements against written criteria. It takes account of evidence in four main areas: the existence of personnel security policies, processes, and procedures; the implementation of a personnel security programme; the consistency of personnel security; and the effectiveness of the policies that are in place. The overall assessment is expressed on a six-point maturity scale: Innocent/Aware/Developing/Competent/Effective/Excellent.[2]

Organisations that have been rated on the PSMM maturity scale can benchmark their maturity against peer organisations – always supposing the results are shared. Benchmarking is a mixed blessing, however. On the plus side, it gives organisations an external criterion to aim for and spurs them on to raise their game. On the downside, benchmarking can encourage complacency if organisations think their work is done once they have reached their target benchmark. It can also cultivate uniformity and stifle innovation if everyone aspires to conform with the same benchmark.

Reaching a benchmark level of maturity begs the question of which level to reach. Should all organisations aim to be 'Excellent', or would a lower level of maturity be appropriate for, say, a small business or an organisation that does not face unusual security risks? What is the right level of personnel security maturity for a technology start-up or a chain of supermarkets? There is no simple answer. An organisation must determine its target level of maturity after considering, among other things, its current defences, the nature and severity of its insider risk, the behaviour of its peers and competitors, and the feasibility and affordability of strengthening security. In some sectors, a regulator or an insurer will have views on how good an organisation's security should be. Regardless of external pressures, an organisation's leadership should ask itself the question: Why would we *not* want to be excellent?

Maturity models have worked well for managing safety risks in safety-critical industries such as offshore energy production, aviation, mining, and petrochemicals.[3] They enable organisations to understand the adequacy of their safety culture by assessing their compliance with specific elements. Improvement plans can be implemented in areas identified as weaker than average. As noted previously, though, what works for safety might not work as well for security. What is more, no security maturity model could provide a template that would work equally well for all organisations, regardless of their size, history, location, and nature of business.

The language of 'maturity' is not always ideal in the context of security. As we saw in Chapter 1, security risks are dynamic and adaptive. To keep pace with the evolving risks, security must itself be dynamic and adaptive. However, the notion of 'maturity' might seem to imply that security has a stable end-state, and that efforts to improve it can wind down once those sunlit uplands have been reached. That is not the case. 'Maturity' has – or *should* have – a more limited meaning in this context. To describe a personnel security system as 'mature' means that it contains the necessary components, and that those components are functioning adequately. It does *not* mean that the system will remain adequate, or that the organisation can relax.

The PSMM methodology is well grounded in evidence, and it has been applied in organisations of many different types. It is also one of the few models designed specifically for personnel security. There are one or two others.

US government maturity models

The US government's Cybersecurity and Infrastructure Security Agency (CISA) has published a comparable scheme for assessing the adequacy of personnel security systems, which it refers to as insider risk management programs. The CISA Insider Risk Management Program Evaluation (IRMPE) framework shows how to build an effective personnel security system. It includes a self-assessment tool for evaluating maturity against benchmarks based on US government policies and research by Carnegie Mellon University. The key components of the IRMPE are consistent with those of the simple threefold model and the PSMM described above. They include: prevention, detection and response infrastructure; training and awareness; data collection and analysis; policies and procedures; protection of civil liberties and privacy rights; communication of insider events; insider incident response planning; confidential reporting channels; oversight of compliance and system effectiveness; and positive incentives.[4]

The US government's National Insider Threat Task Force (NITTF) has published an Insider Threat Program Maturity Framework, which is designed to enhance standards of personnel security in US government departments and agencies.[5] The framework consists of 19 elements under six main headings: Senior Official/Insider Threat Program Leadership; Program Personnel; Employee Training and Awareness; Access to Information; Monitoring User Activity; and Information Integration, Analysis, and Response. Each of the elements represents a capability or attribute that should be present in a mature personnel security system. Again, the parallels with the PSMM and the simple threefold model should be clear.

The cyber framework bonanza

Personnel security has relatively few established models or standards against which to assess the adequacy of an organisation's personnel security. This situation contrasts with physical security, where there are published technical standards for most types of security hardware and systems. It contrasts even more starkly with cyber security, which enjoys a multitude of competing models, standards, frameworks, and accreditation schemes. You can pay your money and take your choice. Two of the most widely used cyber schemes are the cyber security framework produced by the US National Institute of Standards and Technology (NIST) and the ISO/IEC 27001 information security management standard from the International Organization for Standardization.

Most cyber security frameworks pay scant attention to the human dimension, and they are not recommended as tools for dealing with insider risk. An honourable exception is the Cyber Insider Risk Mitigation (CIRM) maturity matrix.[6] As its name suggests, the CIRM describes what good looks like in relation to the insider risk element of cyber security. It covers broadly the same areas as the simple threefold model and PSMM, including governance, assets, risk, and culture, but with a focus on the implications for cyber security.

Subsuming personnel security within a cyber security model is not a good strategy. As we have seen, insider risk is far broader in scope than its impact on cyber systems and data. A personnel security model should encompass the full breadth and diversity of insider risk. While it is true that insiders conduct cyber attacks, making personnel security a necessary part of cyber security, it is equally true that insiders use cyber methods to conduct insider attacks, making cyber security a necessary part of personnel security. Cyber security and personnel security are interdependent but different. By the way, the personnel security models described in this chapter should not be confused with the many taxonomies for categorising insider behaviour which were described in Chapter 8. Insider behaviour is the problem; personnel security is the solution.

What does good look like?

The simple threefold model in Table 10.1 shows the building blocks of a personnel security system. But it does not describe the qualities which those components must possess, both individually and collectively, to make an *effective* personnel security system. The essential or highly desirable qualities for any personnel security system include the following:

1. *Risk-based.* Security is based on a solid understanding of current and emerging risks. Protective measures are applied proportionately, according to the risk.
2. *System-based.* Protection is provided by an integrated system of interlocking measures and capabilities. It does not rely on single point solutions.
3. *Evidence-based.* The design and operation of the system are based on empirical evidence about the nature of the risks and the effects of security interventions.
4. *Well-governed.* The governance arrangements provide clarity about accountability, responsibility, and authority.
5. *Holistic.* The security system is integrated and takes full account of the interdependencies between personnel, physical, and cyber security.
6. *Dynamic and adaptive.* The security adapts rapidly to changes in the risks.
7. *Regularly tested.* The system is regularly tested to ensure that it works as expected and to identify emerging weak spots.

8. *Demonstrably effective*. There is credible evidence that the system performs effectively in mitigating risks.
9. *Legally compliant and culturally acceptable*. All security measures must be lawful. They should also be acceptable within the prevailing culture of the organisation and wider society.
10. *Reliable*. The system should operate reliably and consistently under a range of reasonably foreseeable circumstances.

Thus, a first stab at judging an organisation's personnel security system would involve assessing the extent to which it includes the components of the threefold model and possesses the qualities listed above.

US personnel security authorities take a broadly similar approach, though with different terminology. The ideal personnel security system (which they call insider threat program) is described as 'an integrated, proactive, risk-based mission enabler that makes its organization operationally resilient against insider threats'.[7]

Measuring insider risk

As stated earlier, personnel security should be based on the risk. Ideally, the risk would be described by meaningful metrics. The right metrics would help an organisation to understand the severity of the risk it is facing, which would in turn inform its decisions about how much money and effort to invest in personnel security.

The measurement of insider risk is harder than it might seem, and relatively few meaningful metrics are available. Some of the metrics that *are* used are of limited value or potentially misleading. Even if good metrics were available, analysts may not have access to enough data to estimate the risk with any confidence. Therefore, a reasonable working hypothesis is that no organisation knows the precise size of its insider risk or how to measure it accurately. There might be honourable exceptions to this rule, but they appear to be few and far between.

Why is insider risk hard to measure? The first thing to say is that it is not unique in this regard: other big security risks, including terrorism and hostile foreign state activity, are also hard to measure in accurate and meaningful ways.

As we saw in Chapter 1, security risks have three components: threat, vulnerability, and impact. A risk measurement scheme must capture all three. Threat is hard to measure because the intentions and capabilities of covert threat actors are generally uncertain or unknown. Threat actors cover their tracks and adapt their plans opportunistically.

Vulnerability should be easier to assess because it is a property of the organisation's own capabilities. However, the relevance of any particular vulnerability will depend on the nature of the threat. For instance, a building would be vulnerable to a vehicle bomb if it had no hostile vehicle mitigation (otherwise known as bollards). But that vulnerability would not matter greatly if there was no credible threat of terrorists attacking it with vehicle bombs.

Impact, the third component of security risk, is multidimensional and not easily reduced to a simple numerical metric like financial cost. As we have seen, insider attacks can have multiple effects, including deaths, psychological injuries, reputational damage, business disruption, loss of intellectual property, loss of trust, damage to infrastructure, remediation costs, lawsuits, and tighter regulation. These effects materialise over different timescales, with some taking months or years to unfold. Distilling all these disparate factors into a quantitative metric of risk is inherently difficult. Indeed, some might argue that the problem is intractable. Qualitative metrics are often the best solution.

Organisational versus individual risk

When thinking about metrics of insider risk, it is important to distinguish between metrics that describe the overall insider risk to the organisation and metrics that describe the insider risk associated with an individual. The two are distinctly different but often conflated. Leaders and regulators want to know how much insider risk an organisation is exposed to; personnel security specialists want to know how much insider risk a potential recruit or existing employee might present. Organisational risk could be high because of a large number of slightly risky people or because of a small number of high-risk people.

What makes a good metric?

For any metric to be described as 'good', it must possess two cardinal qualities – namely, validity and reliability. A valid metric is one that measures what it is supposed to measure. In other words, there is a statistical relationship, based on empirical evidence, linking the metric with the relevant real-world outcomes. For example, a valid metric of insider risk would be proven to correlate with observable insider actions and outcomes. If a metric is not valid then it is not meaningful, because it is measuring something else.

A reliable metric is one that produces consistent results. If a metric is unreliable, you will get different results from repeatedly measuring the same thing. A metric may be reliable but not valid, or valid but unreliable. A clock that is set to the wrong time will display a metric that is reliable but not valid. Behavioural scientists and psychologists know they must prove empirically that their metrics are valid and reliable before using them in publishable scientific studies.[8] However, policymakers, security practitioners, and vendors are sometimes guilty of quoting metrics whose validity and reliability are untested and dubious.

Some argue that security metrics should have a third quality, which is that they should be actionable.[9] Metrics are tools designed for a purpose. They should tell a story that helps to improve the quality of decisions and hold people to account. To achieve that end, metrics must be communicated to the right audiences in the right language. A common device for presenting security and risk metrics is the dashboard: a one-page summary of the most important metrics, using graphics to convey information in an easily digestible form. Every organisation should have an insider risk dashboard.

Bad metrics of insider risk

Not all metrics are good metrics. A commonly quoted metric in personnel security is the number of insider cases or incidents. Some organisations interpret this as a metric of insider risk, drawing the dubious conclusion that a small number of cases signifies a low level of insider risk. Not so. The number of cases or incidents is unlikely to be a valid metric of risk, because many insider cases are not detected or not reported. In previous chapters we saw examples of insiders remaining undiscovered for years, while multiple red flags were ignored. The number of cases is more likely to reflect the organisation's ability to discover and record insiders. The less you look, the less you find, and the lower the risk appears to be. An organisation with minimal personnel security might delude itself that its insider risk was low and therefore its meagre defences were sufficient. The organisation might in fact be sitting on a large insider risk of which it was blissfully unaware, with low case numbers giving false assurance.

The number of insider cases could also be an unreliable metric if an organisation does not consistently apply precise definitions of what constitutes an 'incident' or 'case'. Many known insider cases started as HR or compliance cases because of the individual's problematic behaviour or performance. The point at which an HR or compliance case becomes an insider case is often ambiguous. If the criteria move around then so too will the numbers.

Other commonly used metrics in personnel security include the numbers of security breaches, HR cases, disciplinary cases, or dismissals. These metrics can provide useful indicators of organisational risk because of the presumed links between certain types of behaviour and insider risk. In technical terms, they are said to have face validity. Proving that they are valid metrics would require evidence of significant correlations or causal relationships between the behaviours and insider risk. Such evidence is seldom presented.

Lagging and leading indicators of risk

A disadvantage of metrics like numbers of security breaches, disciplinary cases, or dismissals is that they are lagging indicators of risk. In other words, they represent the symptoms of insider behaviour, rather than its early-stage precursors. Metrics like these can help organisations to recognise that they have an insider problem, but they are less helpful in the quest to prevent potential insiders from progressing along the path to harmful action.

We also need *leading* indicators that capture the early warning signs. At the organisational level, leading indicators would convey factors that are known or suspected to foster insider risk – for example, indicators of workforce disgruntlement, bad management, or toxic culture. The so-called key risk indicators (KRIs) that appear on many organisational risk dashboards should be leading indicators. At the individual level, leading indicators would capture signs that an individual is on the path to becoming an active insider and therefore some intervention would be advisable.

Absolute versus relative metrics

Policymakers sometimes assume that the only good metrics are quantitative metrics, i.e. metrics that measure the defined variable in numerical units like pounds sterling or frequency per year. This assumption is wrong. As we have seen, the complex nature of security risks makes them inherently difficult to reduce to simple numerical metrics like cost or fatalities. Threat, vulnerability, and impact do not easily lend themselves to meaningful quantitative metrics that can be compared in absolute terms. However, it is relatively straightforward to compare them in *relative* terms. So, for example, it makes little sense to say that a particular type of security risk (Risk A) has a score of 25.7 units, or that it is 47 per cent bigger than another type of risk (Risk B), or that Risk A has increased by 19 per cent over the past year. But it is possible to judge with some confidence that Risk A is worse than Risk B, and worse than it was last year. Similarly, it is unclear how reputational risk could be measured objectively on an absolute numerical scale, whereas it would be possible to judge that Event A would be reputationally more damaging than Event B.

A familiar example of a relative metric is the UK government's national threat level system for describing the current overall threat to the UK from terrorism. The assessment is expressed on a five-point scale: Low, Moderate, Substantial, Severe, or Critical. The five levels embody relative judgements – Severe is worse than Substantial, and so on – but they do not attempt to quantify those judgements on an absolute numerical scale.

There is nothing wrong with relative metrics of security risk. They are more likely to be valid than pseudo-quantitative metrics that start with subjective judgements which are converted into numbers, lending them spurious precision. A cardinal principle of measurement is that it is better to measure the right things imperfectly than measure the wrong things precisely. The wrong metrics might be easier to quantify, but they can be dangerously misleading if they lack validity, such as falsely implying that a low number of insider cases means the insider risk is small. We should not allow the 'tyranny of metrics' to leave us counting things that can be counted while ignoring the things that really matter.[10]

Better metrics of insider risk

While no metric could adequately encapsulate the totality of insider risk in a single number, it is possible to construct metrics that describe specific components or indicators of insider risk, and which collectively help to build a picture of the overall risk to the organisation.

Examples of possible metrics are shown in Table 10.2. Their validity depends on the organisation having a relatively mature personnel security system that includes effective mechanisms for discovering and responding to insider risk. As noted before, metrics like the numbers of incidents or investigations are meaningless if the organisation is incapable of detecting incidents or running investigations.

Defining metrics is easier than collecting the data needed to evaluate them. Some of the underlying variables are not amenable to simple counting and must instead be derived from professional judgements. A further problem is that key events, such as known insider attacks, happen infrequently. The situation is different with cyber security, where many of the relevant events, such as attempted external intrusions into the network, happen with higher frequency and can be captured automatically, producing rich streams of near real-time data.

Measuring personnel security

Similar considerations apply when measuring the effectiveness of personnel security. Some of the more meaningful metrics will of necessity be qualitative, relative, and based on professional judgements rather than numerical measurements.

The ultimate metric of security effectiveness is that nothing happens. A more nuanced metric would capture the extent to which the personnel security defences reduce insider risk to a tolerable level. The inherent difficulties of measuring insider risk make it hard to construct such a metric. However, there are metrics that capture particular aspects of how well a personnel security system is functioning. Examples are shown in Table 10.3.

Two of the most revealing metrics are time to detect and time to respond. The faster the detection, the better the chances of preventing harm, and the faster the response, the less time the insider will have to cause more harm. Insider fraud cases are found to take an average of 12 months to be detected, and the longer they remain undiscovered the greater the loss.[11] The average time to contain cyber security incidents involving insiders is about 85 days, according to a 2022 global study.[12]

Effectiveness versus performance

When attempting to measure personnel security it is important to distinguish between effectiveness and performance. An effective personnel security system is one that mitigates the risk. Performance, on the other hand, is about speed, cost, efficiency, and targets.

Performance obviously matters, but effectiveness should matter more. Regrettably, organisations are often inclined to care more about the performance of their personnel security, which is also easier to quantify. Prioritising performance over effectiveness is not a good idea, given the potential tension between the two. The easiest way to boost the performance metrics for personnel security would be to reduce its effectiveness. Examples of possible metrics of personnel security performance and costs are shown in Table 10.4.

Table 10.2 Some metrics of insider risk

METRIC	DESCRIPTION	UNITS
Risk events	Events in which insider risk has materialised in some defined way (e.g. violation of security procedures, fraud, misconduct).	Number, frequency per year, trend
Near misses	Events involving precursors to insider actions or attempted insider actions that did not succeed (e.g. failed unauthorised attempts to access sensitive data).	Number, frequency per year, trend
Incidents	Defined incidents in which specified types of insider action took place.	Number, frequency per year, trend
Bullying and violence	Incidents of workplace bullying or violence.	Number, frequency per year, trend
Misconduct cases	Number of cases of defined misconduct.	Number, trend
Misconduct dismissals	Number of people dismissed for misconduct.	Number, % of all departures, trend
Investigations	The number of people under investigation for misconduct or possible insider action.	Number, % of workforce, trend
Expenses fraud	Cases of deliberate abuse or misuse of expenses.	Number, trend
Assets at risk	The assets that would be harmed if an insider risk were to materialise (e.g. data, IP, money, reputation, people, infrastructure). The greater the composite value of assets, the greater the potential impact and hence the greater the risk, other things being equal.	Qualitative or estimated financial equivalents
Impact	The harm that would result from defined types of insider attack. The greater the potential impact, the greater the risk, other things being equal.	Relative scale (Low to High)
Vulnerability	Composite measure of the gaps and weaknesses in protective security that could be exploited by insiders. The greater the vulnerability, the greater the insider risk, other things being equal.	Relative scale (Low to High)
Threat	The intentions and capabilities of insiders and associated external threat actors. The greater the threat, the greater the risk, other things being equal.	Relative scale (Low to High)
Awareness of personnel security	Percentage of workforce tested in survey who have a reasonably accurate understanding of personnel security procedures and their responsibilities.	% of people surveyed
Understanding of risk	The extent to which there are effective functions for discovering, assessing, and communicating insider risk. Lack of understanding is an indicator of heightened risk.	Relative scale (Low to High)
Drivers of risk	Organisational and external factors that are likely to increase insider risk, e.g. bad management, reorganisations, conflict, economic hardship.	Qualitative

Table 10.3 Some metrics of personnel security effectiveness

METRIC	DESCRIPTION	UNITS
Time to detect	The average time taken to detect an insider behaviour of interest (e.g. breach of security policies) or an insider incident (e.g. insider theft of data).	Days, weeks, or months
Time to respond	The average time taken to respond once an insider incident or behaviour of interest has been detected.	Days, weeks, or months
Time to resolve	The average time taken to resolve insider cases, from initial detection to final resolution (e.g. dismissal).	Days, weeks, or months
Time to remove access	The average time taken to remove a person's legitimate access once it is no longer required or permitted.	Days, weeks, or months
Time to restore	The time taken to repair or restore assets and resume normal functioning following an insider incident.	Days, weeks, or months
False positive rate	The number of detections that turn out to be false alarms (i.e. the suspected individual was innocent).	Number or %
False negative rate	The number of genuine incidents or actions that are not detected. This would have to be an estimate based on indirect evidence.	Number or %
Response rate	The percentage of reports raising concerns about potential insider risk that are responded to.	% of reports
Compliance	The extent to which security rules and policies are followed.	% complied with

Table 10.4 Possible metrics of personnel security performance and costs

METRIC	DESCRIPTION	UNITS
Volumes	Numbers of security clearances or renewals completed.	Number per year
Turnaround times	Average time taken to process a security clearance or renewal.	Days, weeks, or months
Backlogs	Numbers of cases awaiting security clearance or renewal.	Number, trend
User satisfaction	Satisfaction of users of personnel security, distinguishing between management and the people subjected to the process, who may have different views.	Qualitative
Complaints and appeals	Cases where the process or its outcome has led to objections or formal appeals.	Number or % of cases
Unit cost	Average cost of completing one security clearance or renewal.	Estimated cost
Relative cost	Average cost per person of pre-trust personnel security as a percentage of the total cost of recruiting and onboarding a new person.	% of recruitment cost
Relative cost	Cost of personnel security as a percentage of total expenditure on security and compliance.	% of security cost
Relative cost	Cost of personnel security as a percentage of the organisation's total revenue or annual budget.	% of budget
Relative cost	Cost of personnel security as a percentage of the estimated maximum foreseeable loss from insider action.	% of potential loss

The value of personnel security

A third type of metric would throw light on how personnel security brings positive value to an organisation.[13] With careful thought, it should be possible to construct meaningful metrics that describe beneficial effects attributable to personnel security. Examples might include:

- Reduction in business disruption
- Reduction in compliance costs and regulatory overheads
- Reduction in insurance costs
- Increase in customer or stakeholder confidence
- Competitive or commercial advantage
- Improved organisational resilience
- Wider organisational benefits from greater trust (e.g. motivation and retention)
- Cost avoided from fewer insider incidents

Metrics like these are likely to be qualitative, although some could be given an estimated financial value based on professional judgements.

Measuring trust

Trust is the universal currency of insider risk and personnel security, so an ideal way of measuring both would be by measuring trust. Again, this is easier said than done. Trust is not a simple commodity that can be reduced to a simple metric. Efforts to measure trust and trustworthiness have focused on proxy indicators that can be assessed more easily.

Measuring trust in individuals

As usual, there are good metrics and bad metrics. An example of a bad proxy indicator of individual trustworthiness would be a person's spoken accent or the way they dress. Apart from having little or no objective relationship with trustworthiness, cosmetic indicators like these are prone to manipulation. Fraudsters know how to exploit these empty trust signals to dupe their victims into trusting them. We are all vulnerable to being swayed in this way.

Trust involves subjective judgements about the expected behaviours of other people in particular contexts. As we saw in Chapter 5, we might judge someone to be trustworthy if they appear to possess a combination of benign intentions, integrity, competence, and consistency. Accordingly, a crude proxy indicator of trustworthiness might be documentary evidence that a person has the qualifications and experience needed to be competent at the job. Observing a person's behaviour over time could improve our assessment of their competence and consistency. However, benign intentions and integrity are not so readily observable. Researchers have investigated these attributes by assessing people's expectations in a range of different situations, either by asking them questions or by inferring their expectations from observations of their behaviour.[14]

Interpersonal trust has been studied extensively in laboratory experiments. The archetypal experiment involves a so-called trust game, in which two players make trust-based decisions about sharing small amounts of money according to simple rules. For instance, Player A has £10 and can either keep it or give some to Player B, who can invest the money and make it grow.

Player B can either keep the windfall or return some or all of the money to Player A. The amount that Player A transfers to Player B is considered to be a measure of A's trust in B, while the proportion that B returns to A is considered to be a measure of B's trustworthiness.[15] Experiments of this sort have supplied insights into the nature of trust between strangers. For example, they show that people often do trust strangers, despite having no prior evidence of their trustworthiness, with both players benefitting from the resulting cooperation. If the game is repeated, each player learns more about the other person's trustworthiness and adjusts their choices accordingly. However, experiments like these are far removed from complex real-life decisions with serious consequences.

Technology has been proffered as a better way of assessing trust and trustworthiness in individuals. Techniques such as eye tracking, skin response, voice analysis, brain imaging, EEG, and automated analysis of non-verbal cues have been investigated as ways of measuring trust.[16] Advocates argue that these methods are more objective than asking people questions, because people do not always tell the truth. Just like the polygraph, however, they depend on the assumption that complex internal constructs like deception and trust reliably generate meaningful physiological and behavioural signals which the technology detects. This assumption is questionable. That said, these techniques do appear to be quite good at detecting when two people seriously *dis*trust one another.

Reputation

How do we assess the trustworthiness of a person or an organisation if we have had no direct experience of interacting with them? A proxy measure of trustworthiness is reputation, which reflects the collective view of others in a highly compressed form. The reputation of a person or an organisation reduces complexity by helping us to predict their likely actions. In the absence of direct personal experience or specific evidence, reputation may be the only guide. Good reputations are famously hard to win and easy to destroy, in much the same way as trust. However, reputation relies on indirect evidence from third parties, which may not be reliable. It is also vulnerable to manipulation. Businesses spend fortunes on burnishing their reputations through advertising and public relations. The main mechanism by which businesses communicate their reputation is their brand.[17] Brands are immensely valuable. Indeed, a company's brand reputation could be its most valuable asset, which is why the reputational damage caused by a major insider attack could matter more than the immediate financial loss or other impacts.

Measuring trust in organisations and nations

In addition to branding, organisations have other ways of trying to appear trustworthy. One way is by publishing performance metrics that emphasise their competence and consistency. Another way is to embrace transparency.

Performance metrics can be presented as proxies for organisational trustworthiness. The expectation is that people will trust an organisation more if they believe it to be competent and consistent in what it does. Performance metrics are a mixed blessing, however. The public know all too well that government and commercial organisations have a history of massaging their metrics to create a positive impression. They are rightly suspicious. Worse, a single-minded focus on performance targets can drive perverse outcomes, as organisations become incentivised to do things that boost their figures or hit arbitrary targets. When the targets become more important than the trustworthiness they were meant to signal, people are left feeling distrustful.

Transparency is another dubious proxy for trustworthiness. The implicit assumption is that people may be more inclined to trust organisations that expose their inner workings to the world, thereby demonstrating that they have nothing to hide. The presumed virtues of transparency are expressed in adages like 'sunshine is the best disinfectant' – a cliché that is both literally and metaphorically untrue. Transparency is hard to dislike as a principle, but whether it does much to make organisations more trustworthy is doubtful.

Organisations publish huge volumes of information in the name of transparency. Journalists and members of the public must sift through it to uncover any awkward evidence of wrong-doing, but the sheer volume can obscure the truth. The published information may not always be reliable because, paradoxically, transparency creates incentives for impression management and dishonesty. Organisations may be tempted to massage the information if they think they will get away with it. And even when published information does reveal wrongdoing, there may be no real consequences for the wrongdoers. Transparency may not be all it is cracked up to be.

Transparency is no better at denoting trustworthiness in our personal relationships. As the philosopher Onora O'Neill pointed out, we can sustain close personal relationships without continually burdening each other with full disclosure of our financial dealings, love lives, and health problems. O'Neill argues compellingly that the best way to build trust is not through transparency, but by bearing down on lies and deception. The real enemy of trust, she says, is deception, and transparency is not an effective remedy for deception.[18] Most personnel security professionals would probably agree that deception is a problem to which transparency is not the solution.

Nation states also have problems with trust. People do not always trust governments or public institutions, and some governments do not trust the people. Public trust in institutions is measured with structured surveys, allowing comparisons across sectors and nations. One of the best known is the Edelman Trust Barometer, which is conducted annually in 28 countries.[19] The 2022 report for the UK showed a marked drop in public trust in government, with a majority (59 per cent) of British people believing their politicians were more likely to lie and mislead the public, and a larger majority (66 per cent) believing that the actions of politicians were under-mining democracy.[20] One positive note was the finding that in the UK and many other countries, people's trust in their employer was higher and rising. Employers are now among the most trusted institutions.

In contrast to the UK, the Edelman survey results for China showed a high and increasing level of trust, with a remarkable 91 per cent of people saying they trusted their government.[21] Of course, the results for China and the UK may not be directly comparable. The Chinese state regards many of its own population as potential insiders and it subjects them to surveillance through its social credit system. The Chinese authorities harvest vast quantities of data from mobile devices, social media, websites, financial institutions, transport networks, and so on, and use it to derive rolling assessments of the trustworthiness of individuals and institutions. These trustworthiness scores have real-world consequences, affecting people's employment prospects, creditworthiness, and freedom to travel, among other things. Opinions differ as to whether the social credit system represents a credible effort to increase levels of trust within Chinese society, or a dystopian abuse of technology to exert control over the population.[22]

Discussion points

- What is the quickest way to assess an organisation's personnel security?
- Would you know if your organisation had an active insider?
- How easy would it be to evade personnel security in your organisation?

- Does personnel security provide value for money?
- Is measuring insider risk an intractable problem?
- How can we measure the value of resilience?

Notes

1 NPSA. (2023). www.npsa.gov.uk/personnel-security-maturity-model

2 Ibid.

3 Foster, P. and Hoult, S. (2013). The safety journey: Using a safety maturity model for safety planning and assurance in the UK coal mining industry. *Minerals*, 3: 59–72.

4 Theis, M. (2022). *CISA's IRMPE Self-Assessment: An Overview*. Carnegie Mellon University. https://apps.dtic.mil/sti/pdfs/AD1158794.pdf

5 NITTF. (2023). *Insider Threat Program Maturity Framework*. www.dni.gov/files/NCSC/documents/nittf/20181024_NITTF_MaturityFramework_web.pdf

6 Hurran, C. (2016). Cyber insider risk mitigation maturity matrix. *Cyber Security Review*, Autumn 2016. www.cybersecurity-review.com/articles/cyber-insider-risk-mitigation-maturity-matrix/

7 Costa, D. and Miller, S. (2020). *From Mitigating Insider Threats to Managing Insider Risk*. Carnegie Mellon University. https://apps.dtic.mil/sti/pdfs/AD1110416.pdf

8 Bateson, M. and Martin, P. (2021). *Measuring Behaviour: An Introductory Guide*. 4th edn. Cambridge: Cambridge University Press.

9 Campbell, G. K. (2015). *Measuring and Communicating Security's Value: A Compendium of Metrics for Enterprise Protection*. Amsterdam: Elsevier.

10 Muller, J. Z. (2018). *The Tyranny of Metrics*. Princeton NJ: Princeton University Press.

11 ACFE. (2022). *Occupational Fraud 2022: A Report to the Nations*. https://legacy.acfe.com/report-to-the-nations/2022/

12 Ponemon. (2022). *2022 Cost of Insider Threats Global Report*. Ponemon Institute.

13 Campbell, G. K. (2015). *Measuring and Communicating Security's Value. A Compendium of Metrics for Enterprise Protection*. Amsterdam: Elsevier.

14 Bauer, P. C. and Freitag, M. (2017). Measuring trust. In *The Oxford Handbook of Social and Political Trust*, ed. by E. M. Uslaner. Oxford: Oxford University Press.

15 See, for example: Kohn, M. (2008). *Trust. Self-interest and the Common Good*. Oxford: Oxford University Press; Kumar, A., Caprano, V., and Perc, M. (2020). The evolution of trust and trustworthiness. *Journal of the Royal Society Interface*, 17: 20200491.

16 See, for example: Kok, B. C. and Soh, H. (2020). Trust in robots: Challenges and opportunities. *Current Robotics Reports*, 1: 297–309. https://doi.org/10.1007/s43154-020-00029-y

17 O'Hara, K. (2004). *Trust. From Socrates to Spin*. Cambridge: Icon Books.

18 O'Neill, O. (2002). *A Question of Trust*. Cambridge: Cambridge University Press.

19 Edelman. (2022). *Edelman Trust Barometer 2022: Global Report*. www.edelman.com/trust/2022-trust-barometer

20 Ibid.

21 Ibid.

22 See, for example: Kobie, N. (2019). The complicated truth about China's social credit system. *Wired*, 7 June 2019. www.wired.co.uk/article/china-social-credit-system-explained; Yang, Z. (2020). China just announced a new social credit law. Here's what it means. *MIT Technology Review*, 22 Nov 2022.

11   Barriers to success

Reader's Guide: This chapter looks at common obstacles to understanding and managing insider risk, including lack of systems thinking and cognitive biases.

Why does personnel security sometimes fall short of what is needed? Many factors can hamper the attainment of good security, including unclear governance, shortage of subject matter experts, and reluctance to share information. Some common barriers to success are listed in Table 11.1. Several, including governance and metrics, have been discussed in previous chapters. This chapter will examine four other problem areas: lack of systems thinking; cognitive biases; shortage of empirical evidence; and unsuitable methodology.

Lack of systems thinking

Insider risk is a systems problem requiring systems solutions, as discussed in Chapter 6. When designing and operating a personnel security system, it helps to think about the distinctive properties of complex adaptive systems, such as interdependence and non-linearity. This means recognising, for example, that security measures do not operate independently of one another, and that well-intended interventions can have unintended consequences. Some organisations reach for point solutions; they are seduced by vendors of magic bullets and rely on a few disparate measures like pre-employment screening and automated monitoring. Security professionals prefer to rely on systems.

Systems thinking aligns naturally with the doctrine of holistic security, which tells us that physical, personnel, and cyber security should be managed collectively. Again, though, there is often a disparity between theory and reality (as depicted in Figure 6.1). Personnel security is usually the worst affected, not least because much of the information needed to understand and manage insider risk resides in other organisational functions.

Systems thinking facilitates a strategic approach to personnel security, which should amount to more than a collection of policies, processes, and technologies. Good personnel security has a strategic purpose, which is to understand and manage insider risk. A more expansive purpose would be to build trust and resilience. As we saw in Chapter 5, high-trust organisations benefit in many other ways besides having less insider risk. The right kind of systems-based personnel security can enable the development of trust and strengthen organisational resilience.

Cognitive biases and psychological predispositions

Why do we make decisions that, with the benefit of hindsight, look wrong? Why do we contemplate the possibility of damaging events but wait until disaster has struck before taking decisive

DOI: 10.4324/9781003329022-14

Table 11.1 Barriers to success

• Shortage of subject matter experts
• Unclear governance
• Rapid turnover of personnel
• Lack of systems thinking
• Inadequate understanding of insider risk
• Insufficient engagement by leaders
• Not treating insider risk as a distinct risk in its own right
• Confusing absence of evidence of threat with evidence of absence of risk
• Psychological predispositions and cognitive biases
• Inadequate sharing of information between functions and between organisations
• Legitimate concerns about proportionality and intrusion
• Unfounded concerns about privacy and confidentiality
• Lack of meaningful metrics
• Focusing on processes rather than outcomes
• Shortage of empirical evidence and data
• Not utilising knowledge from research
• Not learning from experience
• Poor organisational culture and security culture
• Cyber-centricity
• Excessive reliance on pre-employment screening and behavioural controls
• Unsuitable methodology

action? Why do we behave as though low-likelihood/high-impact risks will not materialise on our watch, or that bad things only ever happen one at a time? Why do groups of decision makers coalesce around a faulty judgement and stick to it? If we knew why, we could make protective security better. The answers lie in our psychology.

Biological evolution has equipped us with an array of psychological predispositions and cognitive biases that systematically influence how we perceive the world, make decisions, and respond to novel situations. These predispositions may be thought of as unconscious heuristics, or rules of thumb, which helped our ancestors to survive and thrive in a risky and uncertain world.[1] We would not survive long if our every action required rational, deliberative judgement. Most of our everyday decisions are made quickly and without conscious thought. Our psychological heuristics are remarkably good at helping us to navigate uncertainty at speed. They are evolved capabilities, not design flaws. Nonetheless, our cognitive biases can lead us astray in our current environment of complex risks and technologies. They have been described as predictable deviations from normality.[2]

In the context of protective security, our biases can skew our perception of security risks and impair our ability to manage those risks. The biases play out at both the individual and organisational levels. They pervade the processes by which organisations make strategic judgements, including how seriously to take insider risk. We therefore need to understand these biases and, where possible, offset their less helpful influences. Fortunately, psychologists have been studying them for many years and there is a substantial body of research evidence to draw on.[3] The psychological predispositions and cognitive biases that are most relevant to insider risk and personnel security are outlined below.

Optimism bias

We tend to believe that bad things are less likely to happen to us than to other people. We also find it easier to believe things that we *wish* were true, leading us to expect that everything will turn out well in the end. A degree of optimism is healthy; it reduces anxiety and helps us to persist with difficult tasks and be resilient in the face of adversity. But delusional optimism is dangerous if it stops us from confronting major risks. The danger of not being sufficiently alarmist is illustrated by the 1986 Chernobyl nuclear disaster, in which the reactor operators made a series of bad decisions based on judgements that turned out to be over-optimistic. The disastrous chain of events was reinforced by groupthink and confirmation bias (see below). Optimism bias also appears to have contributed to suboptimal responses by governments to the Covid-19 pandemic.[4] Optimism bias can encourage us to make risky decisions that subsequently look reckless and ignore risks that require action. In protective security, optimism bias can nurture a comforting belief that the risks are tolerably small, even when objectively that is untrue. It can breed a kind of hubris about insider risk as something that only affects other people's organisations.[5]

Base rate bias

When considering how much we should worry about an adverse event, we tend to ignore the general prevalence of such events in the population as a whole (the so-called base rate). Insiders and insider incidents are rare: most people are not active insiders and most early warning signs will be false alarms. If the prevalence in the population is low, then even a highly accurate detection system may generate more false positives than true positives. Suppose, for example, that in a workforce of 1,000 people there are two active insiders – a base rate of 0.2 per cent. Suppose also that the organisation has a fabulously capable automated monitoring system that is '95 per cent accurate', in the sense that it has false positive and false negative rates of only 5 per cent. If the system identifies an individual as an insider, what are the chances that they really *are* an insider? Most people think the answer must be about 95 per cent because the detector is '95 per cent accurate'. But the correct answer is less than 5 per cent, thanks to the base rate being much lower than the false positive rate. In other words, any individual identified by the system as an insider is probably *not* an insider. Plenty of people find that surprising.

Fading affect bias

When bad things happen, we should learn from the experience and modify our behaviour accordingly. However, the negative emotions associated with memories of bad experiences tend to dissipate more rapidly than the positive feelings associated with good experiences. This fading of painful memories helps us to recover and move on. As Shakespeare put it: 'Let us not burden our remembrances with a heaviness that's gone'. But the fading of negative emotions can breed complacency if the bad experience that caused them was the last security incident, and the negative emotions were anxiety and a pressing desire to prevent a recurrence. When those unpleasant feelings dissipate, the window of opportunity for action closes, as people turn their attention to other matters. Fading affect bias helps to explain why organisations often wait until a major security risk materialises before they are galvanised into action.

Illusion of control bias

Leaders and policymakers may be inclined to believe that they have more control over events than is really the case. This illusion of control leads them to expect that their actions will reliably produce the desired results. In practice, however, they pull a policy lever and find that their intervention does not produce the outcome they intended. The illusion of control bias encourages politicians to believe they can control economies, and policymakers to insist that large projects must deliver the exact outcomes they specified in advance. The reasons why we have less control than we imagine derive partly from the nature of complex adaptive systems, as discussed in Chapter 6. Security risks have multiple interdependent causes and they do not behave in linear, predictable ways. People overestimate the power of their actions to determine outcomes, and underestimate the role of chance. When dealing with personnel security, the illusion of control can mislead organisations into believing that they are on top of the problem when they really are not.

Present bias (or future discounting)

We worry more about risks that seem likely to materialise sooner, compared with risks that seem further away in the future. This tendency, which is sometimes described as tackling the crocodile nearest to the canoe, is not irrational. Clearly, we *should* attend to the most pressing near-term risks. But the flipside is paying insufficient attention to risks that seem further away, even when those risks are objectively much bigger and closer than we think, such as climate change. Present bias tilts our attention towards security risks that are higher in likelihood, and therefore more likely to materialise in the near term, and away from risks that are lower in likelihood but much higher in impact. And remember: if you wait long enough, a low-likelihood/high-impact risk will become a *high*-likelihood/high-impact risk.

Availability bias

We tend to overestimate risks that come easily to mind and underestimate risks we find harder to picture. We therefore overestimate risks that are shocking, exotic, or in other ways memorable – the sorts of things that feature in movies and social media. For instance, we instinctively think of sharks as more dangerous than hippopotamuses or stairs, even though hippopotamuses and stairs kill far more people every year than sharks. The same bias can distort our perception of security risks by leading us to overestimate the likelihood of events that we find easy to imagine and underestimate risks that spring less easily to mind because they are mundane or less shocking. A damaging insider attack that unfolds slowly over many months, leaving few overt traces, may be less attention-grabbing than, say, a sudden cyber attack that penetrates network defences. If so, we might underestimate the insider risk and overestimate the more easily imagined cyber risk.

Confirmation bias

Personnel security must be dynamic and adaptive to keep up with the changing risk, as we saw in Chapter 1. We must be able to revise our judgements quickly when new information arrives. However, confirmation bias can get in the way. We tend to pay more attention to information that supports our existing opinions, while discounting information that contradicts them. Consequently, our preconceptions and prejudices become entrenched. In a recruitment interview, for example, confirmation bias can reinforce the interviewer's initial impressions and make them less likely to notice later signs to the contrary.

Fundamental attribution bias

We are inclined to attribute the behaviour of individuals to their personal characteristics, while overlooking the external factors that are central to shaping behaviour (see Chapter 4). An excessive focus on internal factors can impede our ability to understand and manage insider risk. If we believe that insider risk is largely a product of people's psychological characteristics, we might overlook other influences for which the organisation is responsible, such as unethical leadership, incompetent management, interpersonal conflict, stressful working conditions, or toxic culture. The fundamental attribution bias helps to sustain the misleading 'rotten apple' metaphor by attributing everything bad to the individual.

Groupthink

We humans are social animals, attuned to our relationships with others. We are predisposed to follow the pack and fall in with the dominant view of our group, even if we suspect that view might be wrong. In a group of leaders or experts, opinions may be swayed by the loudest or most senior voice, making it hard for contrary views to be heard. A group of several people might effectively have only one opinion, although that opinion seems weightier because it has emerged from a group. Groupthink has famously contributed to many decisions that turned out to be disastrous. It affects personnel security by enabling faulty judgements to hold sway.

Hindsight bias

One of the uglier cognitive biases is our habit of being wise after the event. Hindsight bias is our tendency to perceive events as being more predictable after they have occurred. When things go wrong, we blame others for making bad decisions, even if they were reasonable decisions based on what was known at the time. Hindsight bias leads us to misremember our own misjudgements, so that when something unexpected happens we believe we were expecting it all along. We unconsciously edit our memories to make us think we would have made the right decision. Hindsight bias encourages a culture of risk aversion and over-cautious bureaucracy, as people realise that they will be judged in hindsight by the outcome of their decisions, not their quality. They become incentivised to pursue 'safe' options that would be less vulnerable to criticism if something went wrong, rather than doing the right thing according to their professional judgement.

Loss aversion

We are psychologically more sensitive to potential losses than to potential gains of the same magnitude. We would rather avoid a small loss of time, money, or self-esteem than make a larger gain. When making risk judgements that involve a trade-off between potential benefits, such as improvements in security, and losses (cost and disruption), we lean towards the avoidance of loss. Loss aversion inclines organisations towards avoiding the costs of better security. At the individual level, it makes us more mindful of the time and effort required by everyday security procedures, and less mindful of their benefits. Our aversion to these small costs can make us less willing to comply with the rules. Our heightened sensitivity to negative events also affects the nature of trust. Adverse information about an individual's trustworthiness has a bigger influence than positive information, which helps to explain why trust is so easily broken.

Sunk-cost bias

We tend to persist with a course of action in which we have invested a lot of time, effort, self-esteem, or money, even when objective evidence suggests we should cut our losses and stop.

We know we are in a hole, but we keep digging. Sunk-cost bias could be seen as a type of loss aversion: an unconscious desire to avoid the feeling of loss that would result from abandoning a course of action to which we have been heavily committed. In protective security, sunk-cost bias can impel organisations to persist with cherished policies and practices that are demonstrably faulty.

Risk compensation

We tend to take more risks when we think we are better protected. For example, cyclists take more risks when wearing a helmet and drivers drive faster when wearing a seatbelt.[6] Similarly, IT users tend to behave more recklessly online if they think the system is fully protected. Working in a high-security organisation where everyone has been vetted might encourage people to feel too relaxed about insider risk. Risk compensation can dilute the protective effects of security.

Countering cognitive biases

This litany of cognitive biases might seem alarming, and there are more where they came from. All is not lost, though, because practical remedies are available. So, having admired the problem, we can turn now to solutions.

A good place to start is being aware of such biases and consciously considering their potential effects when making important decisions. Awareness can be reinforced through structured training. Researchers have devised courses to counter the effects of certain cognitive biases in professional fields including medicine and the law. Trials have shown that some interventions can lead to measurable reductions in confirmation bias and fundamental attribution bias. However, the overall evidence for the effectiveness of anti-bias training is mixed.[7]

Having raised awareness, some straightforward techniques can help to reduce or offset the effects. These are outlined in the next sections.

Checking key assumptions and quality of evidence

Before making an important decision, it is a good discipline to articulate the key assumptions on which the judgement is based and review the quality of the evidence behind those assumptions. Some assumptions will be more solidly grounded in evidence than others.

What If? analysis, alternative hypotheses, and Devil's advocacy

Various techniques can help to counteract groupthink and confirmation bias by forcing decision makers to consider alternative possibilities before pressing ahead. What If? analysis involves deliberately changing one or more of the key assumptions and then reviewing the possible consequences. The alternative hypotheses technique involves consciously exploring the implications of other hypotheses. In Devil's advocacy, one member of the group is asked to challenge the prevailing view and propose a different one. (In centuries past, Devil's advocacy was the role of the professional 'fool', whose job it was to challenge their monarch or emperor in ways that no one else would dare.)

Lessons-learned sessions

A good practice – though one that is more often talked about than done – is the lessons-learned session that should follow every significant incident. As its name suggests, the purpose is to learn lessons from experience and apply them to future practice. The willingness to do this may be limited in organisations that prefer to gloss over awkward truths rather than risk damaging their reputation. There is an important difference between *learning* lessons and *identifying* them. Learning means identifying the relevant lessons and then – crucially – changing behaviour accordingly. Governments often promise to 'learn lessons' after bad things happen, but get no further than identifying them. A rigorous process for learning lessons and embedding them in operating procedures can help to counter the effects of fading affect bias and availability bias. The best strategy is to learn lessons from other people's experiences. As Otto von Bismarck said, wise people learn from the mistakes of others.

Delphi method

The Delphi method is a structured technique for collecting and distilling the views of experts.[8] It works by asking each expert to submit their views individually and anonymously via an online questionnaire. The views are collated centrally, and a summary is fed back to the experts in an anonymised form. The process is repeated over two or three iterations, with the aim of reaching a consensus and revealing any major areas of disagreement. The Delphi method dilutes the distorting effects of groupthink that emerge when self-confident experts gather in person to debate.

Premortem

The premortem is designed to rescue an organisation from irrevocably committing itself to a faulty decision fuelled by optimism bias, confirmation bias, and groupthink. Just before the big decision is finally signed off, the decision makers are asked to write an imaginary future history of what happened when their decision turned out to have been disastrously wrong. The premortem imposes a brief pause during which the decision makers must examine the possibility that they might be mistaken.

Sleep

Inadequate sleep has debilitating effects on our cognitive abilities.[9] Fatigue makes us even more susceptible to cognitive biases, as tired minds find it harder to analyse problems rationally. One of the cheapest ways to reduce the unhelpful effects of cognitive biases is to get enough sleep. The well-slept mind is better able to employ rational thinking to overcome the distorting effects of unconscious biases. Tiredness is a likely risk indicator for unwitting insider actions, as it makes people error-prone and impatient. The corrosive influence of inadequate sleep is another reason why organisations should treat their people humanely by not persistently overworking or stressing them.

Automated decision-support tools

Another approach to countering cognitive biases would be using automated decision-support tools designed specifically for that purpose. In principle, it should be possible to build AI-enabled systems that guide and prompt human decision makers in ways that systematically offset the effects of optimism bias, groupthink, and so on. The tools could embody techniques like those outlined above – for example, by prompting decision makers to review their assumptions and consider alternative hypotheses before taking the next step. They would guide people to think rationally about problems and evidence before rushing to judgement. To use the terminology of Daniel Kahneman, these tools could help us to engage our system 2 (slow) thinking to augment our instinctive system 1 (fast) thinking.[10]

Using AI to counter human bias would be an interesting twist, because AI is often criticised for being biased and unexplainable. Bias does creep in to automated systems when algorithms and training data incorporate pre-existing human biases. The behaviour of AI systems is also unexplainable, in the sense that no engineer can describe the exact process by which the system arrives at a specific output. But in fairness to the machines, we humans are also riddled with biases and our behavioural outputs are unexplainable, in that no one could specify the precise chain of neurobiological and psychological processes by which an individual arrives at a decision or behaves in a certain way.

Shortage of empirical evidence

The design and operation of a personnel security system should be guided by empirical evidence about the nature of the insider risk and the effects of the defensive measures. Unfortunately, evidence is often incomplete or of poor quality, forcing security practitioners to rely on their experience and professional judgement. In some cases, the evidence is absent; in other cases, there is some evidence, but it derives from poorly designed research with small sample sizes or artificial settings that do not resemble real-world environments. The shortage of evidence is apparent with some technologies that purport to detect insider risk.

There are grounds for optimism, however. The body of knowledge about insider risk and personnel security is growing. For example, US government-funded institutions have been compiling and analysing databases of known insider cases, while behavioural scientists have been investigating the nature and origins of insider risk.[11] Even so, many basic questions remain only partially answered. Some areas of relative ignorance that call out for more research are listed in Table 11.2.

Table 11.2 Areas of relative ignorance requiring more research

- How can we measure a person's trustworthiness?
- How can we improve the effectiveness of pre-employment screening?
- How can we measure trust within an organisation?
- What are the best ways to build trust within an organisation?
- What organisational and environmental factors have the strongest influence on insider risk?
- What are the best early warning signs of insider risk?
- How could automated systems accurately and reliably detect early warning signs of insider risk?
- Which interventions are most effective in reducing insider risk?
- What can we learn about successful insiders who have *not* been discovered?
- How can we deter or disrupt insiders?
- How can we measure insider risk?
- How can we measure the impact and cost of insider behaviour?
- How can we measure the effectiveness of personnel security?
- How can we measure the benefits and value of personnel security?
- How can we assess the insider risk posed by intelligent machines, and how can we protect against that risk?

Unsuitable methodology

The use of unsuitable methodology can impair our ability to understand and manage insider risk. One common problem is the misapplication of quantitative risk management methodologies that were developed for other purposes, like managing project and programme risks or insurance risks. We saw one instance of this in Chapter 9, when considering why the concept of risk appetite is inappropriate for security risks. As noted previously, security risks differ in several significant respects from other risks, and quantitative tools that work in other contexts may not be right for security.

Another barrier to success is the excessive application of project and programme management (PPM) methodology. An organisation with a highly developed personnel security system will need to modify it from time to time, but this can be done through normal business processes. Implementing wide-ranging changes may require a more structured approach, which is where PPM can help – provided it is not over-engineered.

Confusingly, the word 'programme' is commonly used in two different ways. When policymakers talk of a programme, they usually mean a set of projects or workstreams to implement business change and deliver specified outcomes. A programme in this sense has a beginning and an end. After it has delivered the desired change, the programme stops and what follows is business as usual. A programme could be a vehicle for building a personnel security system. However, some organisations refer to their steady-state personnel security system as an 'insider risk [or threat] management programme'. In this other sense, the 'programme' is the personnel security system itself, not the programme of work that built it.[12]

Efforts to improve personnel security sometimes stumble because the methodology lacks rigour. More often, however, organisations err in the opposite direction, by applying excessively complicated PPM processes that soak up disproportionate amounts of money, time, and bandwidth. False assurance is sometimes drawn from the sheer rigour of the methodology, rather than the efficacy of its outcomes. Seasoned sceptics may be wary of change programmes that have the word 'transformation' in their name, as history suggests that not all transformation programmes deliver genuinely transformational outcomes. Not everything in life is a programme, and the sledgehammer of PPM methodology should be used judiciously.

Discussion points

- Does your organisation suffer from optimism bias?
- Does your organisation learn from experience?
- Is your organisation doing enough to share information on insider risk, both internally and with other organisations?
- Is your organisation's personnel security based on reliable evidence?
- Is project and programme management methodology underused or overused in your organisation?
- Who and where are the subject matter experts in this field?

Notes

1 Martin, P. (2019). *The Rules of Security: Staying Safe in a Risky World.* Oxford: Oxford University Press.
2 Haselton, M. G., Nettle, D., and Murray, D. R. (2015). The evolution of cognitive bias. In *The Handbook of Evolutionary Psychology*, ed. by D. M. Buss. Hoboken NJ: John Wiley & Sons.

3 See, for example: Slovic, P. (1993). Perceived risk, trust and democracy. *Risk Analysis*, *13*: 675–682; Sharot, T. et al. (2007). Neural mechanisms mediating optimism bias. *Nature*, *450*: 102–105; Kahneman, D. (2011). *Thinking, Fast and Slow*. London: Allen Lane; Croskerry P., Singhal G., and Mamede, S. (2013). Cognitive debiasing 1: Origins of bias and theory of debiasing. *BMJ Quality & Safety*, *22*: ii58–ii64; Acciarini, C., Brunetta, F., and Boccardelli, P. (2021). Cognitive biases and decision-making strategies in times of change: A systematic literature review. *Management Decision*, *59*: 638–652; Berthet, V. (2022). The impact of cognitive biases on professionals' decision-making: A review of four occupational areas. *Frontiers in Psychology*, *12*: 802439.

4 Pascual-Leone, A., Cattaneo, G., and Macià, D. (2021). Beware of optimism bias in the context of the COVID-19 pandemic. *Annals of Neurology*, *89*: 423–425.

5 Bunn, M. and Sagan, S. D. (2016). A worst practices guide to insider threats. In *Insider Threats*, ed. by M. Bunn and S. D. Sagan. Ithaca NY: Cornell University Press.

6 Gamble, T. and Walker, I. (2016). Wearing a bicycle helmet can increase risk taking and sensation seeking in adults. *Psychological Science*, *27*: 289–294.

7 See, for example: CIPD. (2015). *A Head for Hiring: The Behavioural Science of Recruitment and Selection*. London: CIPD; Sellier, A.-L., Scopelliti, I., and Morewedge, C. K. (2019). Debiasing training improves decision making in the field. *Psychological Science*, *30*: 1371–1379.

8 Linstone, H. A. and Turoff, M. (2011). Delphi: A brief look backward and forward. *Technological Forecasting & Social Change*, *78*: 1712–1719.

9 Martin, P. (2003). *Counting Sheep: The Science and Pleasures of Sleep and Dreams*. London: Flamingo.

10 Kahneman, D. (2011). *Thinking, Fast and Slow*. London: Allen Lane.

11 CERT. (2018). *Common Sense Guide to Mitigating Insider Threats. Sixth edition*. Carnegie Mellon University. https://resources.sei.cmu.edu/asset_files/TechnicalReport/2019_005_001_540647.pdf

12 Trzeciak, R. (2020). *Building an Effective Insider Risk Management Program*. Carnegie Mellon University. https://apps.dtic.mil/sti/pdfs/AD1146176.pdf

Recommended further reading

The go-to source for practitioners is the National Protective Security Authority (NPSA, formerly CPNI), the UK government's national technical authority for personnel and physical protective security. NPSA provides a range of resources through its website www.npsa.gov.uk. See, in particular, the sections on reducing insider risk and the Personnel Security Maturity Model:

www.npsa.gov.uk/reducing-insider-risk

www.npsa.gov.uk/personnel-security-maturity-model

Another excellent source is the series of papers and presentations published by the Software Engineering Institute at Carnegie Mellon University, including:

Common Sense Guide to Mitigating Insider Threats. https://resources.sei.cmu.edu/asset_files/Technical Report/2019_005_001_540647.pdf

Technical Detection Methods for Insider Risk Management. https://apps.dtic.mil/sti/pdfs/AD1110 364.pdf

How to Sniff Out Insider Threats. https://apps.dtic.mil/sti/pdfs/AD1168388.pdf

Five Best Practices to Combat the Insider Threat. https://apps.dtic.mil/sti/pdfs/AD1086798.pdf

Insider Threat or Insider Risk — What Are You Trying to Solve? https://apps.dtic.mil/sti/pdfs/AD1110 414.pdf

Building Out an Insider Threat Program. https://apps.dtic.mil/sti/pdfs/AD1115839.pdf

Other recommended books and papers:

Bunn, M. and Sagan, S. D. (2016). A worst practices guide to insider threats. In *Insider Threats*, ed. by M. Bunn and S. D. Sagan. Ithaca NY: Cornell University Press.

Bunn, M. and Sagan, S. D. (eds.) (2016). *Insider Threats*. Ithaca NY: Cornell University Press.

Campbell, G. K. (2015). *Measuring and Communicating Security's Value. A Compendium of Metrics for Enterprise Protection*. Amsterdam: Elsevier.

Greitzer, F. L. (2019). Insider threats: It's the *HUMAN*, stupid! *NW Cybersecurity Symposium*, April 2019. https://doi.org/10.1145/3332448.3332458

Martin, P. (2019). *The Rules of Security: Staying Safe in a Risky World*. Oxford: Oxford University Press.

O'Neill, O. (2002). *A Question of Trust*. Cambridge: Cambridge University Press.

Wilder, U. M. (2017). The psychology of espionage and leaking in the digital age. *Studies in Intelligence*, *61*: 1–36.

WINS. (2020). *Countering Violent Extremism and Insider Threat in the Nuclear Sector*. Version 2.0. Vienna: WINS. www.wins.org

Glossary of terms

aftercare (also known as in-trust security and ongoing security): The post-recruitment elements of personnel security. Aftercare is intended to mitigate the insider risk from people who have legitimate access to an organisation's assets.

AI: Artificial Intelligence. A catch-all term referring to digital technologies that enable machines to perform complex tasks. Common applications include search engines, facial recognition, language recognition and translation, digital assistants, chatbots, navigation systems, autonomous vehicles, and generative AI. See also ML (machine learning).

active resilience: The ability of an organisation to grow progressively tougher by learning from adversity and becoming better able to cope with future stresses. See also passive resilience.

assurance: The process of actively verifying that all is as it should be.

attack: Hostile action by an insider or other threat actor that is intended to cause harm. An attack represents the materialisation of a security risk.

authentication: The process of verifying the identity and access rights of a person or other entity.

autonomy: The extent to which an insider's illicit actions are self-directed, as opposed to cultivated, directed, or coerced by an external threat actor.

CERT: Computer Emergency Response Team. In the context of personnel security, CERT often refers to the CERT at Carnegie Mellon University Software Engineering Institute, a globally recognised centre of excellence in personnel security.

CHIS: Covert Human Intelligence Source. Also referred to as agent or informant.

CISA: Cybersecurity and Infrastructure Security Agency. US federal agency.

CNI: Critical National Infrastructure. Those facilities, systems, sites, information, people, networks, and processes necessary for a country to function and upon which daily life depends. CNI is a subset of National Infrastructure which, if damaged, would have major impacts on a national scale.

CPNI: Centre for the Protection of National Infrastructure. The UK government's national technical authority for physical and personnel protective security. In 2023, CPNI changed its name to the National Protective Security Authority (NPSA).

critical path: The processes by which an individual develops insider intentions and progresses towards conducting insider actions.

crown jewels: The most valuable physical or virtual assets of an organisation, the loss or damage of which would cause serious harm.

cyber: Related in some way to digital electronic networks or systems.

deterrence communication: Communication designed to deter or otherwise influence threat actors.

digital footprint: All the information that is available online about a person or organisation.

DoD: US Department of Defense.

DoJ: US Department of Justice.

EAP: Employee Assistance Programme. Support to employees who are experiencing personal difficulties such as mental health problems or bereavement.

exit procedures: A set of procedures, implemented when an employee or contractor is leaving an organisation, to ensure that all contractual obligations are met, all equipment and other assets are returned, and all authorised access to virtual and physical assets is terminated.

foundations: The components of a personnel security system that underpin pre-trust and in-trust measures. They include governance, leadership, risk management, and culture.

governance: The way in which an organisation is structured, managed, and led. The key attributes of governance are accountability, responsibility, and authority.

holistic security: Protective security that manages physical, personnel, and cyber security in the round, taking account of their deep interdependencies.

HMG: His Majesty's Government. The UK government.

HR: Human Resources.

HUMINT: Human Intelligence. The process of collecting secret intelligence from covert human sources, as opposed to technical means.

impact: The multidimensional consequences that arise when a security risk materialises. Impact is one of the three elements of risk, along with threat and vulnerability.

insider: A person who exploits, or has the intention to exploit, their legitimate access to an organisation's assets for unauthorised purposes. Alternatively: A person who betrays trust by behaving in potentially harmful ways.

insider risk: The security risk arising from the actions of insiders. Alternatively: The security risk arising from trusting people.

intentionality: The extent to which an insider consciously intends to perform illicit actions that are potentially harmful.

in-trust measures (also known as aftercare or ongoing security): Protective security measures applied *after* a person has been trusted and given access.

IP: Intellectual Property.

IS: Islamic State, also known as Islamic State of Iraq and Syria (ISIS), Islamic State of Iraq and the Levant (ISIL), and Daesh. An Islamist terrorist group.

meta-analysis: A statistical technique used by scientists which combines the results of many different published studies to produce the strongest possible evidence and conclusions about a particular issue.

metric: A specified variable that is measured in some way. Metrics need not be quantitative.

ML: Machine Learning. The ability of computers to learn without being explicitly programmed. ML has been described as a very fast and fancy way of doing statistics.

NCSC: National Cyber Security Centre. The UK government's national technical authority for cyber protective security.

NITTF: National Insider Threat Task Force. The US government federal authority charged with developing a government-wide insider threat program for deterring, detecting, and mitigating insider threats and safeguarding classified information.

NPSA: National Protective Security Authority. The UK government's national technical authority for personnel and physical protective security. Previously known as CPNI, it changed its name in 2023.

NSA: National Security Agency. US intelligence agency.

occupational fraud: The use of one's occupation for personal enrichment through the deliberate misuse or misapplication of the employing organisation's resources or assets.

onboarding: The process of bringing a new employee or contractor into an organisation.

ongoing security: See aftercare.

OSINT: Open-Source Intelligence (as distinct from secret intelligence).

passive resilience: The ability to recover from a setback or disruption and return to normality. See also active resilience.

PERSEREC: Defense Personnel Security Research Center. US federal agency.

personal security: The protection of individuals against threats to their safety or security in their private or professional lives. Not to be confused with personnel security.

personnel security: The system of defensive measures by which an organisation protects itself against insider risk. The system comprises policies, processes, and technologies that can be divided into three categories: pre-trust measures, in-trust measures, and foundations.

pre-trust measures: Protective security measures applied *before* an organisation trusts a person by giving them access. Also referred to as pre-employment screening or due diligence.

process corruption: A type of insider action that involves illegitimately altering an internal process or system to achieve a specific, unauthorised objective.

red teaming: A method of testing the effectiveness of security in which Red Team players simulate the actions of hostile threat actors while Blue Team players attempt to defend the organisation against the attack. Its primary purpose is to uncover gaps or weaknesses in protective security.

resilience (in relation to protective security): The ability to prepare for, absorb, respond to, and recover from disruptive events or attacks, and adapt to new conditions. See also active resilience and passive resilience.

risk: The amount of harm that is likely to arise if no further action is taken. Risk is a product of threat, vulnerability, and impact.

risk assessment: The process of making sense of information about risks and forming a view about their likelihoods and impacts.

risk discovery: The process of uncovering information about new, emerging, or previously overlooked risks.

risk management: The process of discovering and understanding risks, judging their tolerability, and taking action to mitigate them.

risk appetite: The amount of risk an organisation chooses to accept in pursuit of some goal.

risk tolerance: The amount of risk an individual or organisation is willing to tolerate.

SCIF: Sensitive Compartmented Information Facility. A high-security enclave within a building where highly classified information can be read or discussed.

security culture: An organisation's consistent tendency to behave in certain ways with regard to its security.

social engineering: A collection of techniques used by threat actors to deceive and manipulate people into performing actions or revealing information.

threat: The probability (likelihood) that threat actors will make a credible attempt to conduct an attack. Threat is a product of capability and intention.

threat actor: An adversary, enemy, or potential attacker; an individual, group, or other entity with the intention to attack one or more targets.

trust: A psychological state comprising the intention to accept vulnerability based upon positive expectations of the intentions or behaviour of another.

trustworthiness: The extent to which an entity possesses characteristics by which we judge them to be worthy of our trust. The four main components of trustworthiness are benign intentions, integrity, competence, and consistency.

trust propensity: An individual's general psychological predisposition to trust others.

vetting: An ambiguous term that can denote either personnel security in general, or the pre-employment screening elements of personnel security in particular.

volunteer: An insider who spontaneously offers their services to an external threat actor and subsequently acts, to a varying degree, under their direction.

vulnerability: The gaps or weaknesses in the potential victim's protective security defences that could be exploited by threat actors. More precisely, the probability that threat actors would succeed if they were to attempt to attack or breach security.

Index

Milton Keynes UK
Ingram Content Group UK Ltd.
UKHW051537141024
449569UK00028B/1504